Microbial Machines

Microbial Machines

EXPERIMENTS WITH DECENTRALIZED
WASTEWATER TREATMENT AND
REUSE IN INDIA

Kelly D. Alley

UNIVERSITY OF CALIFORNIA PRESS

University of California Press
Oakland, California

© 2023 by Kelly Alley

Library of Congress Cataloging-in-Publication Data

Names: Alley, Kelly D., 1961– author.
Title: Microbial machines : experiments with decentralized wastewater
    treatment and reuse in India / Kelly D. Alley.
Description: Oakland, California : University of California Press, [2023] |
    Includes bibliographical references and index.
Identifiers: LCCN 2023001492 (print) | LCCN 2023001493 (ebook) |
    ISBN 9780520394308 (cloth) | ISBN 9780520394315 (paperback) |
    ISBN 9780520394322 (ebook)
Subjects: LCSH: Sewage—Purification—India—21st century. | Water reuse.
Classification: LCC TD745 .A65 2023 (print) | LCC TD745 (ebook) |
    DDC 628.30954—dc23/eng/20230126
LC record available at https://lccn.loc.gov/2023001492
LC ebook record available at https://lccn.loc.gov/2023001493

32   31   30   29   28   27   26   25   24   23
10   9   8   7   6   5   4   3   2   1

Contents

Illustrations

Preface

My personal journey into wastewater began in the early 1990s when I was investigating sewage treatment infrastructure during the first and second phases of the Ganga Action Plan in India. I was working in the Ganga River basin and investigating how centralized wastewater diversion and treatment systems were working, and also failing. In my first book, I described my many excursions along wastewater drains (nalas) as I explored the complex cultural orientations toward the Ganga as a river goddess and purifier of human sins. Because she was also a receptor for wastewater, I probed the multiple interpretations across science and faith as people engaged with Ganga's waters for spirituality and livelihood as well as urban sanitation.

As time went on, I researched the expanding field of wastewater management in India and grew rather depressed about the slow progress being made in tackling the influx of wastewaters into freshwater systems. The sewage treatment plants were not doing what they were supposed to do, and there were ongoing issues with institutional decision-making that prevented a fuller involvement of public participation in river cleanup. In 2015, I was introduced to a few projects in community-scale wastewater treatment and reuse and became interested in the ways that local participation spurred a greater interest in arresting and digesting waste-water flows. At that time, I had been conversing with my colleagues about the fact that authorities cared much less about end-of-pipe influxes of raw and partially treated wastewater into rivers than they did about fresh-water purification for drinking or about hydropower for electricity along the same river systems. There was little incentive to treat wastewater when it appeared there was nothing to do with it except discharge it into flow-ing drains, streams, and rivers. Even though treated and untreated waste-water was able to augment river flows and recharge groundwater, the value of treating the waters to in-country water quality standards was not being fully realized. However, in projects that led to reuse, the treatment plant owners and operators and the communities of water users appeared to be assigning greater value to wastewater.

As I began to visit small wastewater treatment projects, I found reuse experiments across a diversity of settings and in a variety of configurations. The mood of others was decidedly encouraging. People told me about how they could transform wastewaters into usable waters to meet some of their needs for non-potable water. I learned that wastewater could be consid-ered usable for non-potable uses even when it was rejected for potable uses, and that citizens did not need to focus on potability when expanding their reuse strategies. I found fascinating variations in the ways waters were labeled and used for different purposes.

Anthropologists have a role to play in helping to communicate exper-iments and innovations in appropriate technologies for the new circu-lar economies of water, energy, rare earth materials, and other resources. Anthropological writing can reveal how to do fieldwork and analysis of a technological topic at a practical level. This ethnographic pursuit al-lows for the technical and scientific translation of engineering practices as it lays out the everyday challenges and achievements of individuals and communities.

ACKNOWLEDGMENTS

There are many people who have contributed to this project over the last five years. I am grateful to the Cultural Anthropology Program of the

National Science Foundation for funding the project. I was able to extend the funds over four years and support several students. My collaborations with Nutan Maurya, PhD in anthropology, and Sukanya Das, associate professor at TERI School of Advanced Study in Delhi, were foundational to this project. With Nutan Maurya's consistent efforts to locate new sites and find key informants, we were able to extend our reach to many on-site treatment and reuse projects. She helped immensely in setting up appointments and tracking down new contacts. We spent countless hours discussing these projects and debating ideas. Sukanya Das was involved in a large EU-funded project called Saraswati at the time and lent her experience assessing decentralized sanitation systems, as well as provided background knowledge of on-site programs across India. She identified several of the graduate students who worked on the project and guided the research team in designing the survey instrument and devising sampling procedures.

There were many engineers and academic professors who helped us understand treatment and reuse systems. A. A. Kazmi at IIT-Roorkee was especially helpful in first introducing me to the decentralized projects when we spent two days visiting sites along the Ganga River in the Himalayas. Other academic and professional engineers at IIT–Mumbai, IIT–Chennai, and NEERI were gracious in explaining their systems and inventions. We were also guided by representatives at CURE-India, the CDD, and many other NGOs mentioned in this book.

There were many PhD and MA students who worked on this project. Debaleena Dutta, a graduate student at Auburn University, worked with Sukanya Das to create and administer the survey instrument and design the visual diagrams for that survey. Debaleena set up our initial database of survey data and Karthick Radhakrishnan and Raihan Akhter of TERI helped immensely administering the survey, cleaning the final dataset, and generating descriptive statistics. Ali Krzton, Shiqiang Zou, Pratibha Prakash, Shubhangi Chaddha, and Shreya Annie Mathew created graphs and maps for the project. Jennifer Barr, a PhD student at Emory University, worked on several data management tasks and critiqued the survey and interview questions, contributing insights from her own PhD research on sanitation NGOs in India supervised by Peter Brown. Rachel McKay, an undergraduate student at Auburn University, assisted the survey research for one month, adding a bright spirit to our long days in heavy traffic. Ed Denton, an undergraduate Auburn anthropology student, helped to audit the survey data and crafted an insightful analysis for his capstone paper in sustainability. Tarini Mehta, now a professor at O.P. Jindal University, guided me through judicial constructs and helped me meet advocates and judges during our trip to the Uttarakhand High Court. Evan Berry's religion and climate change project, funded by the Henry Luce Foundation, gave me a platform for presenting and discussing data and analyses over several years. I also benefitted from discussions with colleagues at a conference on nature in the Indian courts at the University of Edinburgh organized by Daniela Berti and Anthony Good. Amita Sinha, Ellis Adams, Katy Sparrow, Neil DeVotta, Sumit Ganguly, Deborah Winslow, Stuart Lane, Vern Scarborough, and my colleagues at Auburn—Paula Bobrowski, Arianne Gaetano, Sweta Byahut, Nanette Chadwick, Carole Zugazaga, Conner Bailey, Kris Shuler, and Wayde Morse—have supported or engaged in discussions with me over the project period and afterwards. I am grateful to the editors, Stacy Eisenstark and Enrique Ochoa-Kaup, and anonymous reviewers at the University of California Press for pushing me to improve this book.

I am emotionally thankful to my close friends and family who endured my absences or comforted me during long periods away from home. Shalini and Arun Shamnath gave me a true home in Delhi, where I could work and rest and enjoy the vibrant company of their other guests and family members. Subhadra Channa, now retired from the Anthropology Department at the University of Delhi, lent her ear on many occasions. Ali and Feroz Baktoo checked in on me while in India to make sure all was well. I am also thankful for the beauty and tranquility of Puerto Aventuras, Mexico, where I found the intellectual space and courage to start this book and see it through. My mother, my children Khayr and Zahra, and Joanie Ferguson listened to more stories than they wanted to hear, giving me the love and support to pursue my writing schedule during COVID lockdowns. Most importantly, I am grateful to the many people we met and interviewed and surveyed during the research, who all communicated their ideas and experiences so generously. This book is dedicated to their efforts and spirit.

Introduction

In bus depots across India's national capital territory of Delhi, public transport buses must be washed before they are sent out for service. In the West Delhi depot, around 45 of the 125 buses lined up in the large parking lot are washed with treated wastewater each day. In Delhi and other metropolitan regions of India, bus depots have relied on groundwater for their water source, but levels are declining markedly. In 2015, state governments and the courts started discussing new measures to curb groundwater use in city parks, construction projects, industries, and bus depots.[1] Around the same time, government leaders, private company chairpersons, and water board officials were developing and experimenting with small-scale treatment systems that could produce usable water from wastewater. In West Delhi, very near the bus depot, a pilot wastewater treatment plant was built by a private company on land housing the city's largest centralized wastewater treatment plant in the locality of Keshopur.[2] In July 2015, Delhi's chief minister, Arvind Kejriwal, well known for his proposals to increase piped water to all households in Delhi, presided over the opening ceremony for this pilot treatment project. Kejriwal took a long sip of the treated water from this pilot plant to draw media attention to treated wastewater and to emphasize its usability. When the

1

Delhi government ordered all bus depots to stop drawing groundwater in 2017 and use treated wastewater for all their cleaning activities, the Delhi Jal Board (Delhi Water Board) constructed a connector pipeline from the pilot project to the Keshopur bus depot across the street and initiated one of the first experiments in wastewater reuse for the city.

One of my first visits to a project involving reuse of treated wastewater was to the Keshopur bus depot. My research colleague and I sought interviews with the depot staff after hearing that they were using treated wastewater supplied by the small pilot project across the street. On the day of our unscheduled visit, we hoped to at least talk to a few people at the gate, to learn about what was occurring inside the depot. We were happy to find that the reception official at the depot was willing to lead us to an office where several staff members including managers, accountants, and supervisors were gathered. After explaining our research interest, we began the interview in a conversational format, with several staff members answering our questions separately and together. Our conversations then segued into discussions on the depot's varied water supplies, the costs for each supply and historical details on getting the pipeline established from the pilot project across the street. During these discussions, I noticed that the depot staff were generating locational understandings of wastewater and reuse and describing their knowledge of the qualities of the waters supplied to them. They were basing their understanding on daily contact and usage and defining different water supplies in relation to water purification, sewage treatment infrastructures, and the microbial reactions occurring within the latter. They were describing the trace metals, substances, and pathogens in treated wastewater. It became clear to me during our interviews that seeing and smelling water qualities and using the treated water for a specific purpose were behaviors that supported the pilot wastewater treatment project across the street. While research on wastewater reuse has emphasized the disgust or yuck factor, wherein treated water is considered repugnant, unusable, and even harmful, it appeared to me that these employees were breaking through the disgust factor and creating knowledges and situations in which wastewater could be valued as a resource. They were voicing their perception that the treated wastewater from the small pilot project was of better quality than the water they received in tanker trucks from the large, centralized treatment plant.

Their experiences touch on insights and challenges in the emerging field of wastewater reuse. In water-stressed regions such as parts of India, the southwestern United States, eastern China, Israel and Arab countries, Namibia, Singapore, and Australia, communities are looking for new water sources to meet ongoing demands and to adapt to changing water cycles and climate change. Wastewater reuse now appears attractive as a less explored but potentially beneficial option. Treated wastewater can provide a water supply for human needs and ecosystems. This book supports the emerging interest in wastewater reuse by describing human engagements with treatment and recycling across several states within India. These innovative projects display variations in technologies, water budgets, and small and large infrastructures.

Yet wastewater reuse poses challenges across the spectrum of human cultural practices and machine functions.[3] These challenges underscore the fact that wastewater is an undervalued resource; on the world stage, treated wastewater accounts for barely 3 percent of water used worldwide. While in many countries, centralized wastewater treatment systems have extensively piped sewerage networks, an accoutrement of pumping stations, bioreactors, filtration and disinfection devices, and ancillary equipment such as backup generators, in India most facilities are deficient in one respect or another and 70 percent of wastewater runs untreated into surface and groundwater. These deficiencies have pushed authorities and concerned citizens to look for other ways to procure water. Authorities and concerned citizens are finding that the most promising way to address both the challenges of water scarcity and the deficiencies of centralized systems is to experiment with decentralized wastewater treatment machines and optimize them to produce reusable water. Across a diverse set of cases, I found that individuals and communities were willing to use grades of treated wastewater when they were directly treating or managing the treatment of these waters and using the reused water for specific purposes. I argue that decentralized experimentation leads to greater acceptance of wastewater reuse.

This argument builds upon a growing body of literature regarding human cultural attitudes and perceptions about reusing wastewater. Public acceptability is a significant problem in the United States, Australia, and other highly industrialized countries. Fielding, Dolnicar, and Schultz found that acceptance of recycled water decreases as human contact with it increases.[4] A community's disgust surrounding wastewater may prevent the expansion of potential uses and applications of treated water. Taking a more optimistic approach, Scruggs has argued that public acceptance of potable reuse is possible but depends on the history of water scarcity, citizen experience with drought and water reuse, community size, the way a project is introduced and by whom, communication strategies, and trust in the officials and entities introducing a project.[5] Members of businesses and communities are identifying and labeling gray, black, and reuse waters. As Barnes has argued for irrigation water in Egypt and Walsh has explained for conceptions of groundwater in Mexico, water is not simply water, but becomes different waters over time and space.[6] Wastewaters are similarly defined as plural and differentiated.

By reusing wastewater, additional water is added to the supply chain to increase on-site availability for communities and businesses. To get to on-site reuse, communities and businesses experimenting with wastewater treatment systems struggle with scale when treating the water to a reliable standard. These communities and businesspersons experimenting with reuse have questions: How much wastewater is needed to reach optimal treatment with an on-site machine? Can a decentralized or modular unit achieve the effluent standards assigned to centralized systems? When I visited the experimental community of Auroville in the state of Tamil Nadu, where community members were deeply engaged with experiments in sustainable architecture, water, and energy, I was able to talk in detail with the director of Auroville's Center for Scientific Research. As he explained, scale is critical:

> We know that in Auroville. We knew that we had to take care of our own energy requirements. The same applies and will have to be done with our own recycling of wastewater. We cannot expect the government to do it for you [us]. We will have to come back to a model which decentralizes it. The thing is how far you decentralize. There is an optimum. If you do it on an individual scale at the household, it doesn't work. We found that out. It is too costly, too complex, too many things. So you have to come back in a cluster design. We have to do that. This is a role that we have to work on. This is a road we have to work on in the future. The tech is less of a problem, high tech or natural. If you come down to a sizeable cluster or quantity of waste-

water that you can maintain, fantastic. There awareness becomes important. You come back to water consciousness. No spoiling, making sure that
everything works, repairing taps that leak. Nothing that the government is
going to take care of. Huge effort. The way forward!"

The challenges involved with building and sustaining decentralized infrastructures involve optimization, maintenance, and repair. Machine operators articulate the requirements of running and repairing machines and
describe how problems develop when technologies fail to work according
to plan. Engineers relate their methods for adjusting and managing sewage
and its biochemistry and concentration to reach optimal treatment conditions. The bacteria that digest and degrade wastewater need suitable working environments. Managers and supervisors must make sure that bacteria
can thrive and digest biological matter within machine phases. Anaerobic
bacteria thrive without oxygen, while plenty of oxygen must be supplied
to aerobic bacteria. If the right scales are achieved, experimenters hope
that decentralized or on-site treatment systems can avoid the problems
that plague centralized systems: the over-expenditures on long-distance
pipes to carry sewage; the energy-intensive pumping systems; the dilution
of sewage from rainwater and runoff; and the corruption that degenerates
public services and trips up regulation and monitoring.

THE HUMAN-MACHINE-MICROBE PERSPECTIVE

Wastewater reuse is not a new idea, but the integrative study of this activity requires new framing. A new framing must consider the disciplinary
and professional lenses that have informed the wastewater sector, including environmental engineering and public administration. It must contextualize the approach within local and regional water availability using
water science and hydrology. It must investigate the social organizations
of governance and the circular economy using approaches in the social
sciences. The framing must bring microbiology into the purview to consider the role of microbes in wastewater digestion.

   To do this, I create a human-machine-microbe perspective, drawn specifically for Indian histories, politics, and economics but applicable with

modifications to other countries and contexts. I use data collected from wastewater engineers, consultants, designers, operators, community representatives, business managers, and regulators to understand the social and professional activities involved with treating wastewater and running microbial machines. I draw from the understanding of the hydrosocial cycle and from studies of human-machine interactivity to form the theoretical perspective. I contribute to discussions on decentralization and the multilayered arrangements of water governance in India. I focus on treatment and reuse systems in businesses and large institutions and within housing communities, leaving aside the more complicated domain of industrial treatment.

The established notion of the hydrosocial cycle considers wastewater a socio-natural or socially embedded substance. It brings together hydrology, or water science, and the social sciences and directs attention to the ways the society—its key actors and institutions—shape water meanings and uses through infrastructures and technologies. In the hydrosocial cycle considered here, consumption practices turn potable and non-potable water into wastewater and then wastewater is transformed into other waters, and some is reused. Waters are named and labeled at specific moments in this water cycle as they circulate from groundwater to consumption water and then to wastewater, passing through phases of treatment and through the life cycles of technologies. Microbial activities are also described as wastewater moves through treatment machines and is stored for use or discarded.

The interactions that resident groups have with machines are central to these hydrosocial cycles. Interactions may occur in a direct way as they build, operate, and maintain sewage treatment plants (hereafter I will use the acronym STP for sewage treatment plant) or as they interact indirectly through funding, decision-making, and the monitoring of projects. The notion of sociotechnical systems originally developed by Eric Trist, Ken Bamforth, and Fred Emery focused on explaining the hierarchical work design in England's coal mines. This initiated the aim of "joint optimization" between people and technology in engineered work systems and workplaces to create the best technological performance for improving the quality of human life.[7] Optimization is central to the work that wastewater treatment plant operators and communities grapple with today.

Operators and communities also bend the boundaries of "human" and "machine" when integrating parts of an infrastructure. Like Haraway's cyborgs with machinated body parts that help the body function where the organ system has failed, humans help to connect segments of infrastructure through cleaning, machine repair, and the transport of fecal waste and wastewater, at times endangering their bodies and health.[8] Operators of machines are also involved with the microbes that are integral to bioreactors. As Rose has shown, microbes are part of other species' life that humans engage with.[9] Governments include microbes in their visions of national biosecurity.[10]

I shape my perspective on human-machine interactivity by focusing on processes of optimization that involve the behaviors of microbes. Murray-Rust et al. have studied the social networks embedded in automation design by doing ethnographic fieldwork with users, trainers, designers, programmers, and engineers who embed their knowledge into a system.[11] I convey the ways engineers, STP operators, nongovernmental agencies (NGOs), government agencies, and community members view and define optimal states, processes, and outcomes. Optimization is needed to achieve desired water qualities in the out-fluent or the treated water, to achieve a standard of BOD (biological oxygen demand) generally in the range of "10" mg per liter. In some cases, optimization requires surveillance and feedback through biometric and sensor devices.[12] However optimization has a long way to go before treatment machines can be considered intelligent or "smart."[13]

Machine operators must use microbes to optimize wastewater digestion. In the human-machine-microbe perspective, machines made and optimized by humans are not networks but bio-machines. They are bioreactors. Described by one respondent as the "heart of treatment," bioreactors are designed to work by anaerobic or aerobic digestion. The systems use one or both of two kinds of bacteria. These are anaerobic bacteria, which do not need oxygen to eat the food in the wastage, and aerobic bacteria, which require oxygen to consume food and multiply. Bioreactors can be highly mechanized and energy-intensive systems or follow low-energy methods that require less maintenance. Some systems use plants and wetlands and mimic natural processes to perform bioremediation. All bioreactors involve managed or spontaneous processes in which microbiological organisms degrade or transform contaminants to less toxic or nontoxic forms.

MICROBIAL DIGESTIONS AND INFRASTRUCTURES

Infrastructures in this field of wastewater management are configured according to the qualities and quantities of human waste.[14] Contributing to the scholarship and ethnography of physical and material networks,[15] new technologies,[16] megaprojects,[17] energy grids,[18] water and sanitation facilities,[19] and solid and military waste,[20] I add the role of microbes to the analysis of machines and grids.[21] Sanitation infrastructure can be distinguished in terms of two fields of activity, wastewater management and fecal sludge management. I am primarily focusing on wastewater management. Wastewater management aims to capture the liquid waste running from toilets, bathrooms, kitchens, and other places and diverting it through pipes, treatment plants, and drains. The infrastructure for this management includes: (a) toilets; (b) pipes and sewers; (c) open or closed drains (where an open drain is referred to as a *nala* in Hindi or a *raja kaluve* in Kannada); (d) pumping stations; (e) conventional treatment plants with large bioreactors; and (f) small decentralized plants with a variety of bioreactors. The other field of sanitation, fecal sludge management (FSM), removes the waste from toilets and septic tanks, which is highly concentrated and not mixed with much water. It is removed from homes with pumps or carted away in large trucks or lorries.[22] Some solid fecal matter is transported from individual homes by manual scavengers who engage in the demeaning and dangerous tasks of solid removal with hands, wheelbarrows, and small transport devices.[23] In fecal sludge management, the infrastructure consists of dry latrines, toilets, septic tanks, and fecal sludge treatment facilities.

When wastewater flows through settlements and cities, underground pipelines may render the infrastructure invisible for a time, but open drains are a reminder that wastewaters are hard to contain and leach out from all corners of human constructed infrastructures. Once created for storm water drainage, many drains carry large volumes of wastewater. As open or underground, earthen, cement, or brick conduits they direct

wastewater to a treatment plant or to the lowest-lying water body. Myriad drains emerge in the interstices of a city's highway and building infrastructures, and in the degraded spaces between settlements and alongside
roads. Drains discharge into water streams and degrade any fresh water
that may be flowing through tributaries of a river system. These drains can
also contaminate groundwater supplies through seepage into the soil. In
the unsewered, peri-urban areas that lie outside the metro grids, households and large building managers direct their wastewater into streetside
drains, which then run toward the nearest water body or low-lying landscape. Manual scavengers, septic tank owners, and cleaners also dump the
fecal sludge they collect into open drains, as there are very few centralized
treatment plants capable of treating this very concentrated sewage. When
wastewater runs through drains and across terrain, soil moisture, or green
water, is also affected. During monsoon rains, open sewage drains overflow into residential areas and introduce pathogens, heavy metals, and
toxins that are dangerous for public health and well-being.

Machines break down frequently, and it takes time to fix or replace
parts. There is a good amount of sociotechnical tinkering or reworking of
the infrastructure of plumbing and pipes. Bioreactor inventors and operators tinker and rework the microbial communities and residents of housing complexes rethink and upgrade the systems installed by builders.[24]
Biological reactions occur in response to the concentration of sewage
in the bioreactor. The aerobic process requires the introduction of oxygen at specific times and rates, and if electricity to run the aerators quits,
that digestion process can falter. When segments of the sewage transport
system—the pipes and pumping stations—are clogged or broken apart by
flooding or other weather and human disruptions, wastewater makes its
way out of system pathways and diffuses into human settlements or runs
eventually toward the lowest-lying areas on land or in surface water.[25]

DECENTRALIZATION

Small treatment systems manage wastewater in volumes ranging from
fifty kiloliters to one million liters a day and serve communities of twenty-
five to a few hundred families. The daily volumetric capacities of most of

the treatment plants I introduce are between forty and five hundred kilo-liters. By contrast, large conventional systems can treat from ten million liters up to eight hundred million liters per day (termed "mld"), the equiv-alent of from 2.6 million to 211 million gallons per day. In both small and large systems, there are three phases—primary, secondary, and tertiary. Each phase produces a different level of water quality and usability. But almost all the projects I introduce stop short of reaching the highest water quality for use as potable water and are not intended to meet drinking water standards. For those experimenting with these machines, potability is not a criterion for acceptability or for the use or perceived success of a project.

Decentralized projects draw from smaller sewersheds, which is the area of wastewater drainage for a specific community. The sewersheds de-scribed in this book are created by aggregates of wastewater producers in a housing community, a business complex or building, or a university or college.[26] The planning, installation, and operation and maintenance of these facilities create innovations in housing and plumbing design, in the practices of collecting wastewater, and in institutional rules and citizen engagement. This means that housing communities or societies, five-star hotels, and universities are directly involved in generating new gover-nance models within the decentralized sewershed.

The intellectual reasoning for decentralized systems emerged as engi-neers, consultants, and scientists realized the pump and energy require-ments and the periodic ineffectiveness of large, centralized treatment systems. In the early 1980s, a group of innovators of applied technologies for sanitation were brainstorming on ways to create small-scale systems. They formed an organization in Europe called BORDA (Bremen Over-seas Research & Development Association). These organization members reached out to Indian scientists and practitioners who then built a net-work of nongovernmental organizations within India. Over time, BORDA and the network of Indian NGOs, together and independently, developed projects to realize Sustainable Development Goal 6—ensuring access to clean water and sanitation for all—through decentralized sanitation, inte-grated planning, and participative processes.

More recently, the concept of a distributive system has entered pro-fessional and engineering discourses surrounding wastewater.[27] The Water

Environment Federation and the Water Research Foundation have defined distributed systems as "an array of decentralized wastewater technologies . . . in small to mid-sized municipalities, as well as large municipalities, and in new land development projects . . . under a single management entity."[28] Lui et al. have defined a distributed urban water system as a system that collects, treats, and reuses wastewater at multiple wastewater treatment plants and supplements a centralized water supply system with potable water.[29] The projects outlined in this book will contribute new understandings of decentralized systems, as they are articulated by innovators of treatment plants, community and business experimenters, and users of recycled water.

THE CUNNING STATE AND NONCOMPLIANCE

Decision-making and rule structures for sanitation are spread across a number of institutions and communities in India. There are rich bodies of literature in India studies and the social sciences that trace out rule making practices in India. Included in these are studies that focus on the practices of noncompliance with rules and laws that surround large-scale projects with energy, water, and other important resources. High-level government committees within India, including the central government planning commission, the Niti Aayog, have argued that noncompliance limits national progress. Within wastewater management, noncompliance appears in STP functioning. Randeria's notion of the cunning state sheds light on the problem of weak compliance and implementation, which she argues is executed by a callous government disconnected from citizen needs. In her view, the Indian state is neither weak nor strong but able to capitalize on its perceived weakness in order to render itself unaccountable to citizens and institutional rules.[30] The notion of the cunning state is helpful in revealing the contradictory and illusory self-representations of the modern state, but, as Lewis has noted, it overestimates the unity of the Indian government with its diverse state governments and vast bureaucracies and underestimates the commitment to policies when officials see fit to implement them.[31] Roy has argued that the government's "calculated informality"—that is, its ability to turn on informality to achieve certain goals—has a distinct temporal dimension; something which has been tolerated for a long time may be abruptly terminated and demolished or might become regulated and formalized.[32] The state may keep regulations intentionally vague to ensure its own flexibility in future development activities. On the other hand, as Mathur has shown for the National Rural Employment Guarantee Scheme in the state of Uttarakhand, the bureaucracies with their colonial legacies create a maze of paper tigers.[33] In neighboring Pakistan, government paper production connects with cultural practices involving status and networks.[34]

I use these approaches to political culture and governance, while primarily focused on state actors, to inform the behavior of government officials and also the private-sector participants in public-private partnerships and in contracts and consultancies with community members who run their own STPs. State actors are not the only cunning agents. Nonstate actors are central to the collective action of building new decentralized systems. Moreover, since legal mandates, municipal rules, and permit requirements are complicated and difficult for some participants to follow, community members may rely upon construction builders and STP operators to manage their compliance issues as they remain unaware of the many dimensions. Monitoring and compliance are dynamic fields of activity involving legal debate and judicial will. In these government, resident, and business responses, cracks appear, operational inefficiencies develop, microbes die off, and machines break down. The microbe-machine problems must be managed by operations or hidden from regulators. When problems are hidden, then fakery and deception develop, and this turns the study of infrastructure into an investigation of pretension, fabrication, fakery, and discursive fibbing. These micro-actions and practices are also included in my human-machine-microbe perspective.

When decentralization occurs in the field of wastewater management, costs are rolled over to businesses and communities who then become involved in evaluation of treatment machines as well as the waters that are produced from them. The costs of machines and infrastructures are powerful incentives or hindrances to the promotion of treatment and reuse and involve developing strategies for collective action. When Joshi and Shambaugh surveyed tannery owners in Kanpur, a heavily industrialized town in Uttar Pradesh, about their involvement in a common effluent treatment plant (CETP) in the neighborhood of Jajmau, they found that the rule requiring tanners to pay into the common effluent treatment system led to a failure of the experimental project.[35] Tanners reported difficulty making payments into the common pool system, which amounted to Rs. 400,000/month for each company. Instead, tanners bribed government inspectors to avoid complying with any of the standards. While such payment schemes for common pool facilities are difficult to maintain, taxes and incentives to improve plant efficiency may be better at achieving pollution control. But Murty and Kumar found that since most sanitation devices are end-of-the-pipe treatment technologies, tax incentives may also fail to promote the efficient use of machines.[36] Legislation requiring all polluters to follow the same level of compliance (e.g., for large as well as small systems) may also result in inefficiency.

Olson and others have argued that collective action involves appropriate scale, and large groups may not be able to organize around common benefits if the benefits to each member are too small.[37] Chidambaram, when finding differences in collective action around water provision and toilet construction in several slums of Delhi, addressed the problem of low agency around sanitation projects.[38] She found that the infrastructural characteristics of toilets could not be easily converted within the built environment, and such redesigns in plumbing were expensive or illegal. While obstacles may exist in low-income communities that lack household toilets, the cases of decentralized projects show breakthrough innovations in the construction of cluster-based treatment plants and redesign of plumbing for buildings. Infrastructural constraints do not necessarily limit collective action.

In the United States, Urade and Gondane found that the factors preventing a project team from implementing a gray water reuse system included capital cost, maintenance cost, local plumbing codes, local water conservation issues, and complexity of the system.[39] LEED (Leadership in Energy and Environmental Design) credits and the spirit of sustainability may exert positive influence on a design team's decision to implement a gray water reuse system. Scruggs noted that Australia's home water recycling systems, improved agricultural conservation measures, and backup desalination facilities were funded by taxpayers but led to sizeable increases in water rates—by almost 100 percent in some cities.[40] These other country examples alert us to the necessity of accounting for the costs of treating wastewater for reuse, as they are explained by users and those experimenting with projects. The accounting provided in this book will fill a gap in the understanding of how communities and businesses perceive and manage costs for on-site treatment and reuse. Their statements, spread throughout respondent narratives, offer a window into the emerging economics of this circular economy. The recorded costs produce benchmarks along the scale of water costs for residents and businesses. But for the moment, the costs of long-term management of systems and water products remain unreachable since most projects are at an early stage in their evolution.

OBSERVING MACHINES AND HUMAN-MACHINE INTERACTION

This ethnography of human-machine-microbe interactivity takes guidance from Murray-Rust et al.'s notion of entangled ethnography which advocates for observation of both human and nonhuman actors at the intersection of their interactions. I have followed the practices of residents of housing societies, managers of hotels and other businesses, university professors, nongovernmental organization members, consultants, engineers, and government officials in their involvement with machines and machine phases. A researcher must corroborate the information provided in interviews and surveys using other methods to triangulate and verify machine and microbial functions. Such monitoring exercises are important for regulators that are insisting on implementation of the law, but these endeavors are critical to the research process and are needed to make firm assessments about on-the-ground realities. This probing involves returning to specific sites over an extended period; revisiting on-site treatment plants and their engineers and managers; tracking the stepwise implementations of court orders; and witnessing the phases of installation, operation, and maintenance that are so critical to human-machine interfaces.

Emergent on-site treatments and reuses of wastewater may produce changes in human engagements with infrastructure that are not yet visualized or anticipated. During my field research, I often heard the phrases, "We don't really know how to do this"; "we haven't thought of that yet"; or "we are not sure how to approach this problem." I saw these as tentative expressions of figuring things out since everything was in the process of invention and testing. There were innovations and experiments in which some citizens were trying to solve a problem, but the solutions were not fully visualized. In emergent situations, the researchers cannot assume that there is a perceptible consensus on best practices for human-machine-microbial interactions. But in their diversity, experiments are innovative and fascinating, and experimenters are encountering new phenomena. Citizens are grappling with machines and technologies in situations of water stress, financial and infrastructural incapacity, and public health risk.

Approaching machines and human-machine interaction involved many challenges. It was not easy to see a sewage treatment plant or witness its functioning over time. Opportunities to observe moments of machine operation, optimization, and repair had to be cultivated and exploited when possible. Finding where machines were located could be a Sherlock Holmes endeavor. It took perseverance to follow the smells and sights of fetid flows. When starting the process of investigation in many places, I often asked residents where the treatment plants were located within their communities, housing complexes or cities. Many could not give a rough location when the scale was large. "It is somewhere over in that part of town," some said. Or they pointed questionably in another direction. Sometimes the locational recommendations led to one of the system legs or components in the wastewater treatment chain and I was challenged to piece them together. For instance, residents could mistake pumping stations for treatment plants. The water is cleaner there, some would say. Upon tracing the pipelines, I would find that it was a midway point on the long journey through neighborhoods into pipelines and then into drains and maybe treatment plants. Since I had to start from scratch—without clear maps—when tracing out diversion and treatment systems, I had to smell around, as it were, to figure out where the sewage was traveling.

When I got to the point that I needed to ask questions from people officially in the know, then the responses grew more complicated. Generally, a government facility will require formalities for gaining entrance, but we also found this to be the case for smaller onsite projects in institutions, businesses, and housing societies. On the other hand, once permission was granted, the tours and explanations of facilities could be lengthy, and hosts could be generous with their time and knowledge. We were able to take pictures of machine parts after obtaining permission.

I started in 2016 by collecting information on decentralized projects from my colleagues and contacts in India. A graduate student that I included in the original project proposal also helped me find the first projects to visit and investigate. Our first visit was to the Keshopur pilot plant. One of the company's technical staff members gave us a well-organized tour and provided the background on how their treatment plant was envisioned. Their experimentation started with attempts to treat the wastewater at their sugar factory. They tried a few technologies to promote digestion of the concentrated wastewater produced through sugar manufacturing. They eventually arrived at a technological method using earthworms and layers of biomass, followed by membrane and carbon filtration devices. Then they approached the Delhi Jal Board (the Water Board of the city of Delhi) with a proposal. They offered to build a pilot treatment plant on the compound of the Delhi Jal Board's centralized wastewater treatment facilities in West Delhi, using their own funds to get the city administrators interested. After the pilot plant was constructed and started treating wastewater, the company leaders were able to convince the government authorities to take over the funding of operation and maintenance. They developed a public-private partnership with the Delhi Jal Board. The negotiations that followed involved the bus depot managers across the street, and user fees were established. This tour gave us details on how a public-private partnership could be built for decentralized sanitation and started us on a path of more in-depth investigation. As we visited projects across the spectrum of private and public engagement, we collected official materials and documented on-site operations and then probed concepts and plans as much as possible with business and community members, technicians, STP managers, and other company representatives who were willing to share their knowledge and ideas.

Over a four-year period from 2016 through 2019, I visited on my own and with my graduate and postgraduate colleagues over two hundred decentralized projects in Delhi and the National Capital Region (NCR), Metro Bangalore, Metro Chennai, Rishikesh, Dev Prayag, Varanasi, Auroville, and Agra. My research colleagues and I documented STPs at industries, university campuses, hospitals, housing complexes, neighborhoods, airports, malls, and city parks. At many of these locations, we cultivated as key informants those individuals who were involved with planning or operating projects or managing and financing them within a community or business. We tried to build trust by letting them know that we were sincerely interested in their successes and empathetic about the challenges they faced. Key informants were then able to help us expand our pool of interviewees and survey respondents, form focus groups, point us to other experiments and consultants, and lead us to official and company records, NGO reports, and university publications.

After using this snowball sampling method for site visits, we began to think about ways to talk to the broader group of people involved in experimental projects. Communities were not always connected to project operations, as we found with projects described later in the book, such as garden STPs and some STPs in hotels. But we had heard from several conferences and discussions with experts that the greatest number of cases of community involvement could be found in Bangalore. My postdoctoral colleague then visited Bangalore on a scoping mission in 2017 and brought back a preliminary list of housing projects compiled by the CDD (Consortium for DEWATS Dissemination Society), a key organization I introduce in chapter 2. We planned to visit those sites the following year.

Many referrals from NGOs such as the CDD and university researchers provided an entry way to visit projects that were closed to the public. When community leaders, project owners, and facility managers agreed to allow us to see projects and to interview participants, we could use the details of our interviews and conversations to build case studies. As we worked in this field of incomplete documents and directories of institutions, housing complexes, businesses, and agencies involved with wastewater machines, snowball referrals continued to form the most productive method. This was especially true when trying to meet residents of Resident Welfare Associations (RWAs) who proved to be central to the functioning of some of the small-scale treatment plants or STPs I introduce in this book.

I conducted extended interviews with officials in the Ministries of Environment, Forests, and Climate Change; Water Resources (now Jal Shakti);

Power; Renewables; Central and State Pollution Control Boards; the sanitation departments of municipalities and other urban municipal councils such as the New Delhi Municipal Corporation (NDMC). Meetings with key government officials and visits to centralized treatment facilities were sometimes difficult to arrange. It was common to find that officers were not in their offices. Phone numbers were difficult to procure to make appointments. Since wastewater treatment plants were in some cases under the purview of court or tribunal orders or were not running properly and thus disguised by managers and operators, these meetings required additional persistence. Many times, when I was ignored, I kept showing up at the relevant offices. Eventually people would get tired of me showing up and would escort me and my colleagues to an informative person, or to a person who led us to another person. When I finally got a person with whom I could have an informative discussion, I was ecstatic. We could deliberate on the problems and their dimensions, with the pauses and silences of frustration punctuating our conversations. In more lengthy interviews with authorities, project monitors, NGOs, and university teams, I could critically explore what they saw as the institutional constraints and possibilities in this emerging field and learn about their technological innovations and failures.

In between fall of 2017 and the next visit in fall of 2018, one of my graduate students at Auburn worked with my colleague at TERI School of Advanced Studies to create a survey instrument. Before designing parts of the survey related to costs, price sheets were gathered from municipalities and state agencies across India to find out what residents were already paying for government-piped potable and non-potable water and for government-provided tanker water. We used these official price sheets to set the context for what respondents were telling us about charges for water levied by public agencies and private enterprises. We also used water costs taken from municipal cost sheets for potable piped water at volumetric rates and compared them with respondents' statements on their costs. The costs of wastewater treatment in housing societies could then be compared with the costs of more advanced systems in hotels and universities. Relevant information was also gathered from the literature on decentralized systems and stages of wastewater treatment, and we used those materials to create diagrams and demonstrate how treatment works within the survey instrument. Once we were ready to choose survey sites in India, my colleague at TERI coached my students and me on ways to select locations that represented a diverse response pool and a range of income levels and geographic locations in the peri-urban zones. As a field-worker who had never conducted a large survey using a representative sample, I was challenged.

Housing complexes with decentralized wastewater treatment facilities were selected to represent different geographic regions and income groups in Bangalore, the Delhi and National Capital Region (the metro area around the capital of Delhi that spreads into the surrounding states of Haryana and Uttar Pradesh). Geographic regions were identified using census data and city planning documents. Communities within these regions were selected using lists of housing complexes in Bangalore (now Bengaluru) and Delhi that were provided by governmental and nongovernmental agencies and private companies working on decentralized systems. We ended up using a mix of demographic and geographic sampling methods to reach a range of communities representing different income levels and geographical locations. These included: snowball sampling (to meet the most informative respondents within agencies and companies), transect sampling (to get a representation of businesses or housing complexes within a neighborhood or urban area), and respondent-driven sampling (to reach a set of residents in a housing complex in the absence of directories or resident lists). We also contacted representatives of the companies installing and operating STPs to arrange tours of the treatment plants. In some locations, when building companies were handling the operation of the STP they were more suspicious about our motives and did not allow us to enter and see the STP.

Once we were able to reach members of Resident Welfare Associations (RWA) within a housing complex, we could assemble larger groups to conduct the survey. In some places, RWA members organized gatherings so that we could administer the survey to a larger group of residents. They also helped in generating focus group discussions before or after the survey sessions. Some members of our research team were able to attend RWA functions and mill about meeting members willing to talk and answer the survey questions. The survey was eventually administered to 320 residents of housing societies, which captured approximately 5 percent of the resident population in each location. Since there were no lists of residents for each housing complex (at least not available to us), we had to ask people to refer us to others or go door to door requesting their interest in spending about an hour on the survey and discussion. Sometimes we had to find residents who were friends of friends and get our acquaintance to a community started that way. These were far from perfect respondent-driven sampling procedures, given that bias can enter the selection process through the involvement of the recommending party. Since it was difficult to reach a representative random sample in a housing community, we used as many responses as we could gather to construct case studies and think about broader patterns.

The survey was verbally administered in a direct face-to-face interaction to make sure the background information was explained uniformly and accurate responses for complex questions could be elicited. The survey team included one postdoctoral researcher, two PhD students, one MA student, and one undergraduate student, although not all participated in each field visit. Several of the students were fluent in Hindi, which helped substantially in many locations. Although our answers were recorded in English, there were many conversations and interviews conducted in Hindi or with a mixture of Hindi and English. When visiting sites in Bangalore, we used a translator in Kannada and then managed in other locations in that city with Hindi and English.

After collecting sociodemographic and household information including income, family size, dwelling type, water availability, and sources of water supply, the survey gauged awareness and opinions about wastewater reuse; perceptions and preferences about the use of treated wastewater for different alternatives; willingness to pay for four different hypothetical scenarios named A, B, C, and D, which I report below; knowledge, practices, and challenges of conservation and reuses; and the administrator's role in maintenance and regulations. The four distinct scenarios were as follows: Scenario A represented the status quo or situations where wastewater was not treated and not usable for any purpose. Scenario B represented primary wastewater treatment with technologies such as baffled reactors and digesters that produced reuse water for landscaping and car washing. Scenario C represented secondary treatment that produced reuse water for toilet flushing, landscaping, and car washing. Scenario C represented an improved scenario over scenario B and much improved over the status quo Scenario A. In Scenario D, wastewater was treated by primary, secondary, and tertiary methods, and reuse water was suitable for all the uses in scenarios B and C and for potable use. Respondents were shown a pictorial depiction of each scenario to ensure that they understood the options and the general technologies involved. After briefing the respondents about the three hypothetical scenarios beyond the status quo, respondents were asked whether they would be willing to pay for the scenarios B, C, and D in separate questions. Some were already experimenting with scenarios B and C and could state their views on the benefits and challenges within and outside the survey dialogue. The survey asked their maximum willingness to pay for each scenario according to price points or "bids" that we claimed would be add-ons to their monthly water bill. Respondents were requested to take into account their household income before quoting their willingness to pay, or WTP, for the hypothetical scenarios (see Table 1).

For respondents, the most informative part of the survey was the section on scenarios and uses of treated water displayed with pictures. While some respondents were already deeply aware of the phases of treatment and the reuses possible using technologies for each scenario, many others were still in the learning phase about operations and appeared to enjoy learning about what could be produced. In this way, the survey became an awareness-raising and learning device for those just coming into the experimentation process.

During site visits, interviews, and survey work, I tape-recorded many of the conversations that I had with our research team members and our respondents, after obtaining their consent. I rely upon these narratives in both Hindi and English to build the ethnographic descriptions in this book. These tape-recorded discussions provided the ground realities for me as I listened to and organized the responses and conversations. The discussions gave me ideas about how to describe the activities of experimenters by using their exact words in extended quotations. The extended quotations presented in this book contain the details of their projects, and some of the broader discussions on water and wastewater. For me, the survey was more of a qualitative device that could provide answers and generate unpredicted responses. People wanted to talk outside the boundaries

Table 1 Visual Diagram Modified from Survey Instrument of NSF Project

Scenario	Types of Reuse					Quality of Water after Treatment	Type of DEWATS Technology in Use
	Landscaping	Car Washing	Flushing Toilets	Household Cleaning	Drinking		
Scenario A (status quo)	✗	✗	✗	✗	✗	Wastewater is released in the environment **without treatment**	**None**
Scenario B (good)	✓	✓	✗	✗	✗	Wastewater is released in the environment after **primary treatment**	Septic tank and biodigester
Scenario C (better)	✓	✓	✓	✗	✗	Wastewater is released in the environment after **secondary treatment**	Anaerobic baffled reactor or aeration tank
Scenario D (best)	✓	✓	✓	✓	✓	Wastewater is released in the environment after **tertiary treatment**	Planted bed filter + reverse osmosis (RO)

of a question. They wanted to add their own additions and markers. They wanted to tell their stories and ask questions. There were responses such as: "How do you want me to answer that? Is that the right answer that I am giving you?" Many answered "yes" to options that were simply hypothetical. Would you be willing to use treated wastewater for car washing? Toilet flushing? And then some more unique uses such as water for swimming pools or for pets? When looking at those responses in bulk, some interesting patterns are ascertainable, but I have treated the results with flexible thinking. Where the discussions on scenarios were hypothetical, I do not draw any specific conclusions or recommendations on acceptability of using treated wastewater or even the willingness to pay for it. I only assess acceptability through the descriptive vignettes and extended quotations of people experimenting with machines and engaging with active reuses of the treated waters.

Probing legal activity and meeting legal practitioners required different strategies. Advocates were busy, and judges were generally inaccessible. It was considered a conflict of interest to poke into an ongoing case (called "sub-judice") and write about its details. I had to make sure that I did not contribute to the resolution process as I strove to write about it. I had to be a fly on the court wall but also ask critical questions to advocates and probe behind the scenes of court documents. Since compliance can be a performance there are many ways in which court orders are not what they at first claim to be. I needed to follow up with advocates on how and why specific items were added to their pleas.

While interviewing individuals involved with reuse projects, some questions did not get immediate answers. They were met with silences or blank stares, or with what I guessed were fibs and lies. These circumstances forced me to figure out what was considered the normative truth. Was lying about machines and their functioning normative? Constructing my personal truth, I had to make a baseline by seeing things for myself. I needed to see the machines and how and when they were working. I needed to see the outgoing effluent from a treatment plant to know what the treatment process was accomplishing. To do that, I had to snoop around the peripheries of facilities and compounds and trace the signs of sewer lines until I could witness the quality of flows meeting the surface air. I returned several times to check drainage flows, creeping through half abandoned land and dirty, shit-filled spaces. I walked endless miles along nalas that smelled so bad I felt sick. I would cover my mouth with my dupatta and take a few more pictures from different angles. I would ask questions of anyone hanging around nearby. How long have the flows been like this? Do they change? Residents were very knowledgeable: Yes, they are worse at night when no one is around, or they would tell me that wastewater gushes out at specific times of the day. Closer to the STPs, suspicions arose. Who are you and why are you asking this? Do you have permission from the authority to be here? What are you going to do with those pictures?

Confidentiality has been an important consideration when shaping my ethnography. When visiting small-scale projects, there were issues related to the fact that institutions and communities were under court orders to install STPs. Treatment plant managers and operators could be wary of outsiders surreptitiously monitoring their facilities and passing information along to authorized persons within the Pollution Control Board, NGT (National Green Tribunal), or the local water and sewerage boards. On occasion, builders were reluctant to give permission for a general inquiry on these systems when they were trying to meet state and central requirements for construction permits. Working with these constraints, I used individual cases to loosely think about cross-sector situations. For instance, details from many housing societies provided the benchmarks on machine functions, costs, and patterns of governance. Interviews involving managers at two five-star deluxe hotels provided a general idea of the water capacities, technological challenges, and reuse options for large hotels. Details from interviews at ten small hotels in Rishikesh explained situational complexities. Case materials representing different income groups and inventor and consultant groups provide a window on the challenges in procuring water and adding reuse water to the mix.

I have removed the names of buildings, housing complexes, and businesses as much as possible to protect the identities of those who provided important information. Some information was controversial since businesses and housing societies have been under court orders to implement treatment plants and initiate reuse options. In some cases, I have named large hotels because they are already featured in the court orders that I mention and quote from. In some cases, I have mentioned large institutions such as universities and ashrams for the same reasons. Otherwise, the many extended quotations presented in this book are disembodied, without clear identities; they may appear as free-floating perspectives and opinions. But they are representative of real people's voices and positions, and they represent, individually and collectively, the many responses to emerging conditions and challenges. The cases are ethnographic so that the context can be portrayed for each community. I have great respect for those participating in this critical movement to create and run wastewater treatment plants and infrastructures with limited resources. The ethnographic descriptions that follow in this book represent new connections and expressions that fill out this human-machine-microbe perspective.

THE PLAN OF THE BOOK

This ethnography of wastewater management and reuse is a descriptive account of human interactions with technology, science, and engineering. It is divided into seven chapters. In chapter 1, I lay out the cultural and institutional context for this study, pointing to key features in Indian society that uniquely configure the responses brought out through the ethnography. I outline the institutional and legal histories of the regulatory regimes that businesses and communities now respond to as they build decentralized systems. In a cluttered and complicated regulatory landscape, citizens maneuver through the rules produced by government agencies, the courts, and the National Green Tribunal.

In chapter 2, I introduce the inventors of bioreactors and provide descriptions of their machines and the roles microbes play in machine functions. Bioreactors are "the heart" of wastewater treatment. In chapter 3, I begin describing excursions to decentralized sites of wastewater treatment that I have visited alone and with research teams. In chapters 3 through 6, I showcase projects in different stages of evolution and describe how communities interact and respond to machines and infrastructures and to their own water struggles. I portray communities across the income spectrum to explain the disparities in the decentralized management of wastewater. Three examples in chapter 3 convey struggles in low-income communities in peri-urban areas. Chapters 4 and 5 describe projects in middle- and upper-income communities where residents, university members and managers of hotels are experimenting with treatment and reuse of wastewater. The contrasts between languishing projects in chapter 3 and horticultural, partial, off-grid and emerging reuses in chapters 4 and 5 help to amplify the argument that experimentation is a necessary part of building citizen acceptance. The contrasts also point to disparities rooted in income, capacities for collective action, and water availabilities. Responding to new rules requiring decentralized sewage treatment plants (or STPs) at housing societies, hotels, and large institutions such as universities, the experimenters are constantly adjusting their machines and building on-site infrastructures to optimize reuse. Their technologies represent a range of low-energy approaches to advanced multistage systems, each one producing a different grade of usable water. Chapter 6 explores wastewater machines in smaller hotels to highlight the ongoing challenges related to rules and scales that extend beyond the challenges of the acceptability of reusing wastewater. They convey how the costs of running small-scale treatment systems can be a significant hindrance to compliance. In chapter 7, I summarize the features of this human-machine-microbe perspective and redefine the hydrosocial framework for decentralized scales. Experimentation with treatment machines and new forms of decentralization and governance offer possibilities to value wastewater as a resource and provide non-potable water at home and work.

1 Sanitation and Institutional Complexity

The decentralization of wastewater treatment and the on-site reuse of this water require new methods of sharing and owning infrastructure.[1] Communities, governmental and nongovernmental agencies, and private businesses are crafting templates for ownership and management that expand sanitation services and meet demands for new sources of water. Collective action is at the center of these activities. Australia, Israel, and Singapore have centralized reuse experiments involving municipalities and government agencies with regularized fees and taxes. In Israel, government agencies use 90 percent of treated wastewater for irrigation. In Singapore, the government has created an extensive system called NEWater which meets 30 percent of the nation's water demand for high-quality industrial water.[2] In Australia, treated wastewater is injected into groundwater aquifers and then becomes "indirect" reuse water when it is pumped up again.[3] In China, the southwestern United States, and a few Arab countries, treated wastewater is used for irrigation, industries, and for the recharge of groundwater, lakes, and rivers.[4]

In this book, I present a series of case studies to show that decentralized experiments in India are ahead of the global curve of on-site projects that benefit communities, businesses, and large institutions such as universities and colleges. Collective action is pushed by water costs and the scarcity of sources. Institutional and legal mandates motivate collective action within communities, associations, and business groups. In this chapter, I lay out the institutional frameworks and support systems within India that contextualize the case studies. I introduce the laws and water quality standards that communities and businesses respond to.[5] I also explain why spheres of activism have mushroomed from legal foundations in early independent India. Then using participant observations and court orders, I introduce the country's environmental tribunal and give short vignettes of the exchanges between legal activists, government officials, businesses, and justices. At the end of the chapter, I give a short introduction to the rules and regulations that are followed by housing societies and hotels in their management of wastewater.

SANITATION

In previous studies of Indian sanitation, the focus has centered on challenges involved with fecal sludge removal. Also called manual scavenging, fecal sludge removal is a demeaning job forced upon low-caste groups and Dalits. In these jobs, people transport human waste with their hands, and with wheelbarrows and tanker trucks. During colonial India, engineers followed the British model of waste transport and introduced flush toilets to transport wastewaters into stormwater drains. This water-based transport of human waste led to the merging of gray and black waters in sewers and stormwater drains, a merging that complicates efforts to separate and treat concentrated human wastewaters and keep them out of stormwater drains and rainwater runoff. Expanding water transport, the Indian government installed large sewage treatment plants in the 1980s, by upgrading older STPs and establishing new ones in the largest cities. Since then, government programs have expanded these wastewater networks across the country.

These investments in human waste removal have been advanced by two main missions, the Swachh Bharat Abhiyan, or Clean India Mission, which began in late 2014, and the National Mission for Clean Ganga, which started in 2011. Projects funded by these missions have built small- and large-scale sanitation infrastructures and supported public health programs for women and children in their frontline vulnerabilities as keepers of the household and supervisors of hygiene.[6] Although slow and burdened by institutional complexities and weak public financing, they have improved well-being. The Clean India Mission has focused on treatment infrastructure for the fecal sludge removed from septic tanks, buckets, and pit latrines and given financial support for families to build in-house toilets and set up septic disposal services. Treatment facilities have been created for the fecal sludge drawn from septic tanks. This mission has also promulgated media campaigns to end open defecation.[7] The other mission, the National Mission for Clean Ganga, has made significant investments in centralized wastewater treatment plants.[8]

Within the two key missions, the one focusing on fecal sludge management and the other on wastewater management, two professional dispositions have developed with distinctive ways of conceptualizing wastewater and water. Among Clean India Mission activities, waste removal from septic tanks, buckets, and pit latrines is still imbued with the stigma that surrounded the older tasks of night soil collection and manual scavenging that were assigned to lower-caste and Dalit communities. In contrast, wastewater systems are being designed by engineers and consultants, and these elaborate systems of pipes and machines influence the way the public interprets water and wastewater in planning and household maintenance. In previous writing, I have elaborated how the notions of purity, impurity, cleanness, and uncleanness become embedded in discussions about wastewater and its effects on the sacred River Ganga. These terms are part of understanding the conflicting roles of the river as Mother Goddess and receptor of urban and industrial wastewater. Barr has noted that "pollution," "purity," "cleanness," and "dirtiness" are also significant in interpretations and practices of fecal sludge management. In these, purity and pollution are closely tied to caste status.[9] By contrast, I found that sacred purity and impurity were not important in discussions involving decentralized projects. Understandings of the waters produced by sewage treatment machines did not refer to sacred power or caste statuses. The understandings were tied more closely to household, business, and institutional water needs. Discussions were imbued with mechanical metaphors and understandings of microbial digestion. The sociobiological axis of purity and pollution expressed in night soil and fecal sludge removals was receding from the picture as engineers, professionals, and communities conversed about management and reuse. In the process, other distinctions and names for waters have been created.

INSTITUTIONS

The institutional histories involved with the development of sanitation in India follow a number of trajectories. In the remainder of this chapter, I will trace these pathways in the constitution, laws, legal cases in the courts and tribunals, and in specific bodies of rules applied to wastewater management. I begin with the core principles in the Indian Constitution, which direct attention to environmental safeguards. The Directive Principles of State Policy and the Fundamental Duties in the Constitution promote advocacy for the environment. The Directive Principles of State Policy are guidelines for governance that the state is expected to follow when framing policies and passing laws. The Fundamental Duties define the moral obligations of all citizens.[10] Environmental values were inserted into the directive principles and fundamental duties through the controversial Forty-Second Amendment passed in 1976.[11] This Amendment incorporated principles from the Stockholm Declaration issued by the International Conference on Human Environment in 1972. The Stockholm Declaration confirmed the responsibility of each member of society to protect and improve the environment.[12] The Forty-Second Amendment added Article 48A to the Directive Principles of State Policy in chapter 4 of the Constitution.[13] This declared the State's responsibility to protect and improve the environment and safeguard forests and wildlife. It also inserted Article 51A(g) into the Fundamental Duties and stipulated the duty of every citizen to "protect and improve the natural environment including forests, lakes, rivers and wildlife and to have compassion for living creatures."[14] Articles 48A and 51A introduced the obligation of the government and the courts to protect the environment for the people and the nation.

Although environmental protection is included in the Directive Principles of State Policy and the Fundamental Duties of the Indian Constitution, environmental rights are not listed under the justiciable Fundamental

Rights of the Indian Constitution. Hence, they are not directly enforceable. Rather, the directive principles and fundamental duties guides the Supreme Court when it holds that "a citizen has a right to have recourse to [the remedies provided by] Article 32 of the Constitution for removing the pollution of water or air which may be detrimental to the quality of life." In this way, the courts read the Directive Principles into the Fundamental Rights through the remedy of mandamus listed in Article 32. So environmental values are promoted through the commitment, through the mandamus remedy, to rectify the perceived failures of other branches of government. Environmental cases have also benefitted from other procedural advantages attached to the enforcement of constitutional rights.

There are several important sources of law used in addressing wastewater management. The Water (Prevention and Control of Pollution) Act of 1974 ("the Water Act") set out the concerns for water conservation and pollution protection.[15] Parliament adopted minor amendments to the Water Act in 1978 and revised it in 1988 to conform to the provisions of the Environment (Protection) Act of 1986.[16] Although the Constitution determined that water would be a subject in the State List and fall under the purview of the state, the Water Act empowered the Union Government to legislate in the field of state control. That power was affirmed when all states in the Union approved the Act.[17]

The administrative regulation under the Water Act provided for the establishment of a Central Pollution Control Board and, under this, a Board in each state of the Union.[18] These Boards have developed plans for the control and prevention of pollution. The Central Board plans and executes a national program for preventing pollution, carries out research, compiles data, and advises the government on water and air pollution matters. The State Boards implement the Water Act by inspecting industrial and wastewater treatment plants and conducting research on water quality and sewage treatment methods.[19] Both Boards hold the authority to set standards for water quality, air quality, and emissions and effluents from industry and urbanization.[20] The following chapters will refer often to the roles and activities of the state pollution control boards and of the Central Pollution Control Board. Regulatory powers also reside within several ministries within the government of India, namely the Ministry of Environment, Forests and Climate Change, the Ministry of Jal Shakti (formerly the Ministry of Water Resources, River Development and Ganga Rejuvenation), and the Ministry of Drinking Water and Sanitation.

Apart from these constitutional provisions and laws, India's unique form of judicial activism has significantly shaped wastewater management. Within India, public interest or social-action litigation has developed alongside investigative journalism, to spawn concerted public interest in human and animal rights and broad environmental movements. Legal structures and public interest litigation have provided citizens with avenues to engage directly with state and regulatory agencies regarding energy, water, and environmental matters. Public interest cases have tried to hold governments, business leaders, and industrialists accountable to regulations and policies and laws at all levels. This litigation transformed citizens into powerful petitioners and justices into charismatic environmentalists.[21] The courts have heard a range of cases from water pollution, deforestation, and mining to the impacts of extractive industries.

THE REMEDY OF MANDAMUS

It is important to understand the core procedures and remedies that are used in cases focusing on wastewater. The most important remedy is the remedy of mandamus. Early on when justices advocated the need for procedural relaxation to open access to law for all citizens, Justice Bhagwati stressed the judicial remedy of mandamus ["We command" is a judicial remedy in the form of an order from a superior court]. This remedy allowed citizens with few means to seek redress for any public injury arising from breach of public duty or violation of some provision of the Constitution or the law. A citizen could seek to enforce public duty or constitutional or legal provisions and stand for the public interest. For Justice Bharucha, this remedy enacted the mandate of Article 14 that "no one is above the law."[22] He noted, "Public interest litigations have also led to the Courts' pronouncements in pollution and environment matters. I would be the first to concede that these pronouncements have sometimes been a mixed blessing, but we must remember that in these matters, as in so many others, the Court has had to step in because the legislature and the executive had not acted upon their obligation to protect the quality of life."[23] Ongoing legal activism becomes a "continuing mandamus" when citizens or justices directly battle with state agencies and apply monitoring or regulatory procedures on polluting or resource-extractive entities. The courts can also engage in a prolonged issuance of a writ of mandamus to supervise their previous orders.

Petitioners have also been able to stand for the public good using flexible types of petitions such as letters to the court. Additionally, justices have been able to exert influence by introducing cases of their own accord. Justices do this by exercising *suo motu* powers [taken by a court of its own accord, without any request by the parties involved].[24] Justices can use their *suo motu* to intervene directly in the administration of a state-directed or private project.[25] This power has expanded their participation in environmental policy.[26] In the 1970s and 1980s, the Supreme Court heard public interest petitions involving many environmental problems. According to Jariwala, 1998 was the most productive year for judgments.[27] Justices have also appointed commissions to investigate and provide advice on environmental problems.[28] However, outside commissions may be staffed with members of government agencies, who are still able to toe the line of powerful agendas.[29]

The focus on the performance of centralized sewage treatment plants was first addressed by advocate M. C. Mehta, in a case on river pollution filed in 1985. At the time, Mehta was beginning his career as a very active environmental lawyer and many of his cases came before the Supreme Court bench of Justice Kuldip Singh. In this river pollution case, Mehta filed a writ petition charging that government officials, despite the allowances in the legal code, had not taken effective steps to prevent water pollution. While the Ganga pollution cases were heard, the Government of India initiated an environmental scheme called the Ganga Action Plan (GAP). Under the GAP, centralized wastewater diversion and treatment systems were constructed to clean up polluted waterways. The hope was that the treatment plants would be turned over to city municipalities for operation and maintenance, but this never occurred. Instead, the central agencies continued to orchestrate initiatives, especially through the formation of the National Mission for Clean Ganga. That mission then took over all the contracting for large, centralized STPs in the Ganga basin.

MONITORING AND COMPLIANCE

Monitoring and compliance are critical to the success of wastewater treatment and reuse systems. There are water quality standards to abide by and technological prerequisites and best practices. Since the mid-1990s, the new or upgraded centralized wastewater treatment facilities were failing to operate effectively in most locations. As the pollution load increased, the small amounts of treated water were reabsorbed by the untreated wastewater flowing through canals, streams, and rivers. The Supreme Court stepped up its monitoring of sewage management through M. C. Mehta's Ganga case and fined over two hundred industries in the Ganga basin. The courts also penalized the state Pollution Control Boards for false reporting and pressed the Ministry of Environment to streamline its proposals through a less unwieldy set of supervisory committees.[30] In this case and others, monitoring of compliance was necessary to check the functioning and effective use of national assets. Justices were persistent in following up on their previous orders. In cases extending for a long period of time, justices participated in monitoring and regulation, and made up for the work that should have been done by the regulatory agencies, specifically the pollution control boards.

Petitioners and justices have invoked Fundamental Rights, Directive Principles pertaining to environmental quality, and the remedy of mandamus to advance problem solving over crucial water issues. The Fundamental Rights and Directive Principles are broad frameworks of legal justification; the mandamus charge is a broad remedy for executive failures. Using these frameworks, judicial leaders have appointed commissions and reviews to deliberate wastewater problems, find solutions, and check the practices of government agencies and industrial units. This breadth and procedural innovativeness through public interest litigation paved the way for what came next, the creation of the National Green Tribunal.

THE NATIONAL GREEN TRIBUNAL,
GROUNDWATER, AND WASTEWATER

Many cases involving wastewater problems are now heard in the National Green Tribunal (hereafter NGT). The active involvement of the Supreme

Court in the 1990s and its large caseload spurred discussions on the need for an environmental tribunal and the NGT was established in 2010. A few years later, the Supreme Court began to transfer environmental problem solving to the NGT, with the expectation that their legal deliberations would involve a greater degree of scientific assessment and monitoring.[31] This shifted the strategy away from "continuing mandamus" and toward a multi-stakeholder approach. When the Supreme Court started transferring responsibility to the tribunal, the media accused it of "passing the buck." To this, the Court responded, "For us, to monitor week after week, months after months, it is difficult. . . . It [you] can go there [NGT] and if you have a problem you can come back to us." As one informant explained it to me, "When cases are sent by the Supreme Court to the NGT, the NGT is more motivated to hear them."

The Tribunal was set up as a three-to-five-member bench composed of a leading justice, with the title of chairperson, and at least one additional justice, and one to three expert members with knowledge of science and the environment. The expert members were to have a background in environmental assessment or ecological or environmental science.[32] The majority of the Tribunal's cases have dealt with pollution (about 31%) and environmental clearances (about 35%) and the NGT is empowered to investigate environmental harm and issue punishments and penalties. The NGT can also issue relief to petitioners and compensation for damages.

By 2015, the Supreme Court had transferred more than three hundred cases to the NGT. In the 1,130 cases decided by the NGT up to this time, the most frequent plaintiffs were NGOs, social activists, and public-spirited citizens.[33] Gill explained that in the "stakeholder consultative adjudicatory process," stakeholders were brought together alongside the tribunal's scientific justices to elicit the views of those concerned, involving government, scientists, NGOs, and the public.[34]

When M. C. Mehta's Ganga case was transferred to the NGT, the chairperson took over the monitoring of wastewater flows and management. The NGT was also hearing a case on pollution of the Yamuna River petitioned by a citizen activist, Manoj Mishra.[35] In turning attention to wastewater drains in these two cases and others, the NGT aimed to improve data collection, appropriate technology selections, and industrial compliance to water pollution laws. The NGT also started a series of debates on groundwater usage and contamination in the context of petitions filed by citizens.[36] In several cases, the NGT ruled that industries had to curb their use of groundwater, ordering that industries and other large water users had to obtain an NOC, or no objection permit, for their groundwater extraction from the Central Ground Water Authority (CGWA).[37]

THE ROLE OF THE CHAIRPERSON

In October 2016, I was sitting in Court 1 of the National Green Tribunal, as I usually did when I was in Delhi for a period of time. The NGT has expanded enormously since its inception. On any given day, there were more than fifty practicing lawyers present in the court. The courtrooms were filled with advocates, petitioners, and respondents who were constantly shuffling in and out of the room as their cases came up for listing and hearing. With limited seating in the back of the room for observers, I had to strain to hear what people were saying at the front of the busy courtroom

On that week in October, the chairperson of the NGT was leading the bench in a daily assessment of environmental problems. On the days of my visit, the bench was deliberating on the state of the River Ganga's bacterial and toxic pollution load. In these hearings, three separate cases were clubbed together. They were *M. C. Mehta vs. Union of India & Others* (the Ganga case); *Anil Kumar Singhal vs. Union of India & Others*; and *Society for Protection of Environment & Biodiversity & Anr. vs. Union of India & Others*.[38] The petitioners in these cases were trying to hold the institutions involved with environmental management accountable for their responsibilities in water management and pollution control. They wanted the responsible agencies—the pollution control boards, the Ministry of Environment, Forests and Climate Change, and the Ministry of Water Resources—to regulate the myriad small and large industrial units along the riverbanks and in the riverbank towns. The continuous hearings were playing out the older remedy of continuing mandamus. The petitioners and the chairperson could see that the problems were enormous. Expressing his frustration, the chairperson commented on that day:

> The working of the Authorities is pathetic with the passage of time huge amount of public money has been spent by these public authorities, but

while they are only adding to the greater and severe load of pollution to river Ganga and its tributaries. Even today it cannot be informed to us as to how many drains carry industrial effluents, sewage and other discharge to the river Ganga or its tributaries. There are other factors which create serious doubts about the very intent of the State of U.P. or its public authorities in an effort to clean river Ganga in a systematic and proper way.[39]

The bench was addressing the lack of data on the extent of pollution, and as it turned out a few weeks later, on the exact number of polluting industries and entities. The chairperson asked representatives of the main government departments in charge of pollution prevention to list all the sewage drains pouring raw wastewater into the Ganga River and main tributaries. On the days I was present, the chairperson pushed the key agencies to form a joint committee to go out, locate, document, and measure the flow of all the drains from Hardwar to Unnao, as this was the most polluted reach of the river in the states of Uttarakhand and Uttar Pradesh.[40]

The justices and expert members on the bench called on all the concerned government departments to collect data on the drains dumping sewage into the Ganga since none of them could provide up-to-date and comprehensive information on locations and flow volumes. The chairperson instructed in the order on October 16th:

> The hearing of this case is going on day-to-day. The Learned counsel appearing for various Ministries, State Government and State Authorities have prayed for time to seek instructions on the matter in the issue before the Tribunal. We are really disappointed to note that during the course of hearing none of the parties could inform the Tribunal as to how many drains carrying sewage, treated effluent, waste and domestic discharge or others join river Ganga or its main tributary, i.e., East Kali, Kosi, and Ram Ganga falling in Segment "B" of Phase-I in terms of the Judgement of the Tribunal. This area already for our consideration has been extended from Haridwar to Unnao instead of Haridwar to Kanpur. According to the Member Secretary, CPCB who has stated to have personally visited the area, there are 30 main drains which join river Ganga directly and are matter of serious concern from the point of pollution of river Ganga. According to the Learned Counsel appearing for the U.P. Pollution Control Board there are 172 drains out of which 150 drain directly join river Ganga or its main tributaries as aforestated. According to the Learned Counsel appearing for the U.P. Jal Nigam, there are 151 drains out of which 83 drains directly terminate into river Ganga.

However, it is commonly stated that it cannot be identified if the drain in particular carry industrial, domestic or sewage alone. Most of them are carrying mixed effluent and whether they are even connected to STP/CETP or not cannot be determined unless there is proper inspection of the drains. We are primarily concerned with Segment "B" and as already noticed the industrial, domestic and sewage pollution are inseparable and cannot be quantified under their respective head. Thus we have to deal with this issue of pollution of river Ganga collectively in Segment "B" of Phase-I.

The Tribunal cannot come to any definite conclusion till the time the quantum of pollution and quality of pollution is brought before the Tribunal by the concerned authorities. Furthermore, providing of appropriate advise for prevention and control of pollution would directly depend upon the number of drains that are going to river Ganga and its tributaries in Segment "B" of Phase-I. The uncertainty created reflects upon functioning of these authorities who are obliged to perform different statutory duties and maintain record before it could carry out any project of laying of CETP/STP. Officer present from various departments, Ministries and Government are also unable to assist the Tribunal.

We direct that the Member Secretary, CPCB, Chief Engineer of U.P. Jal Nigam, Senior most Chief Environmental Officer of U.P. Pollution Control Board and representative from the Ministry of Water Resources shall personally visit the area falling in Segment "B" of Phase-I in terms of our Judgement and as aforestated. They will identify how many drains join river Ganga or its tributaries and make observations in relation to quantum and quality of effluent that is going to river Ganga or its main tributaries through them. Let this report be submitted before the next date of hearing.[41]

A first reading of this excerpt might give the impression that the government departments were so understaffed that they could not manage the tasks of data collection. But my participant observation in the Tribunal and with members of legal teams and the concerned institutional officers and NGOs yielded insight into these matters of complicity. Central government agencies were pushing forward on the construction of new treatment plants to meet the increasing sewage load in cities, but the development had become deeply rooted in a politics of delay and was twisted by state-center conflicts over funding, jurisdiction and control of water. Critical wastewater treatment infrastructures were being dismembered by this politics of delay and by what Hess called *non-science*, the "lack of data" that had plagued regulatory and monitoring agencies for a long time.[42]

SCOLDING AND DELAY

Another morning started with a lot of joking among the legal teams sitting in the front rows of seats set aside for advocates. At the front, the rows were separated from the elevated bench where the justices sat by two large tables. At these tables, the clerks sorted the files of the day and received additional evidence and documentation as the hearings went on. That morning the lawyers representing the respondents were teasing the petitioner and the primary advocate, M. C. Mehta, about his status and presence. This well-known lawyer had started appearing in the NGT after his cases were transferred from the public interest litigation cell of the Supreme Court. The younger lawyers stood along the edges of the courtroom, pressed between the rows of seats and the courtroom walls. They formed a line of black robes, as advocates chatted with one another about the news of the day. There was the murmur of greetings and conversation before the docket began. Eventually the white-clad bailiffs entered on to the elevated platform at the front of the room and stood waiting behind the chairs of the justices and expert members. All rose. The dignitaries entered and lined up behind the table in front of their chairs. They leaned forward and the bailiffs methodically pushed the chairs toward them. They all sat down. It was a bench of five on that day, so, I thought, something will get done. The chairperson began, "After admission matters, we are starting with M. C. Mehta."[43] There were many problems to remedy and details of bad practices to address. Stacks of cardboard files and books crowded the front section of the room, overflowing with papers that were loosely ordered by sticky tabs.

The day before, the Bench had suspended operations of the sewage engineering agency of the Uttar Pradesh state government, the Jal Nigam, by putting a stay on their use of government funds. Outside the courts, the heat was on to sideline the Jal Nigam, because the central government perceived it to be part of a recalcitrant state government run by an opposition political party. It seemed that the justices' order to suspend operations within the Uttar Pradesh Jal Nigam was loosely connected to the Union government's interest in kicking them out of sewage management projects. In another case pertaining to the status of chickengunya and dengue in Delhi, a justice belted out to the crowd of black robes, "Is there any colony, where complete protections have been taken?" Raising his voice he scolded, "What precautionary measures have been taken by the DDA [Delhi Development Authority] and NDMC [New Delhi Municipal Council]? Name only one colony!" He asked about fumigation and inquired about dirt and standing water. One government servant replied, "It is impossible to ensure no stagnant water in 1,060 unauthorized colonies." The justice scolded the authorities more about lack of implementation. He threatened to take steps against officers who would not ensure implementation of previous orders. "You are charging house and property taxes, no?" the justices asked. "You only know how to make money!"[44]

At this time, the National Mission for Clean Ganga (NMCG, the central government agency overseeing wastewater management across the Ganga basin) was in the process of finalizing a notification that would redefine itself as a "Society." Such a notification would make it a quasi-government entity that is allowed to engage directly with the private sector and foreign government contracts. Since centralized wastewater treatment systems were expensive and required, at that time, significant funds from the World Bank and other foreign banks, the redesignation would give the agency more power to negotiate directly with funders without going through the maze of state bureaucracies. With this designation the mission could exert more power through its central or federal ministries and work around state departments to finance, contract, and implement projects. They could then avoid dealing with state-level institutions in large infrastructure projects. It was not clear on that day how the NGT came to ask about the status of the notification, since the government did not of its own volition offer it up for vetting.[45] The justices were aware that the process was underway in the NMCG to centralize their powers and grow as an entity.

The session of that day revealed that the establishment of these new additional powers was not yet a done deal. The Tribunal had a right to weigh in by examining the language in the notification and asking what it fully meant.[46] Stopping short of accepting the changes, the justices demanded clarification on several items. "The role of each newly defined entity needs clarity with a chart," one justice requested. "Also, the coordination among levels needs to be explained. Finally, execution through the states needs clarity. How will projects be executed through the states?"

On that day, the Pollution Control Boards were also cornered on their "lack of data" situation. Previously, lack of data had been a modus operandi; nobody required data, and nobody asked for it or why it wasn't there. No one demanded new documentation. On this day the discourse focused on developing databases to establish legitimacy. By verifying drain locations and their contents, a proof document could be issued. "Why haven't these drains been identified?" the justice asked. "You need to have lots of information." The departments with wastewater in their purview had created many lists of polluting industries and drains. However, there was a confusion of multiple outdated versions, a situation that in the past had kept specifics cloudy and undecipherable, maintaining the "calculated informality" that Roy so astutely identified.[47] On this day, the Tribunal was demanding updated and corroborated lists.

In the afternoon session on the same day, the bench continued to discuss procedures with the advocates and officials representing the state pollution control board of the state of Uttar Pradesh (UPPCB). The UPPCB officials were asking the justice for an order to require the district magistrate (DM) to close the polluting industrial facilities. They claimed they were powerless to close industries without the support of the DM. The UPPCB functionary said: "We cannot enter an industry without the DM. We have to show cause. It is not possible for a regional officer to get it closed." The justice persisted, "Why can't you? It is in your power." "Sir," the officer continued, "without the help of the DM we cannot disconnect power and other services. We have no real enforcement power."

The discussion then turned back to the number of wastewater drains. "Give us an estimation on the quantum and contents of these drains," the bench continued. The lawyer for the Uttar Pradesh Pollution Control Board shuffled around in his files before claiming that the papers with data on the drains had suddenly gone missing. The justices and other advocates waited and listened in disbelief. My jaw dropped. Then the discussion turned to the policy of "ZLD" (zero liquid discharge of sewage effluent) and whether it was a feasible policy for India. On that same day, there was an intense review of the policy principle of environmental or ecological flows, which stipulated that at least 15 percent of a river's flow should always remain in the river stream. Another discussion covered hazardous waste. The topics were huge, and the discussions touched on the edges of tangled problems. Eventually all were adjourned for the day.

On the next day that the Ganga case was heard in Justice Swatanter's court, advocates and their clients were discussing untreated sewage flowing into the Ganga and Yamuna Rivers while looking at pictures on their phones. While they were looking at the pictures, the chairperson asked, "Are you saying it [the treated wastewater] meets the parameters?" At the same time, the Uttar Pradesh Jal Nigam officials were claiming that three sewage treatment plants (STPs) were old and needed funds. The justice instructed that they take a sample of the treated effluents that were visible in the pictures on their iPhones, together and at a specific time. Then the justices asked about the progress in listing all the polluting industries and the progress on the joint inspection report for all the drains flowing into the Ganga from Hardwar to Unnao. The report was supposed to include the state of sewage, the existing industrial extractions of groundwater, and the disposals of wastewater within drains and on land. The chairperson asked, "Where is the report?" When there was no reply, he followed: "All samples and full report are required by the next date!" He also inquired about whether the industries in question had been granted permission to extract groundwater. One industry owner whose facilities had been closed due to illegal use of groundwater stepped forward to plead that the seal on his industry door be removed. "Kindly please remove it, sir."

In these meandering paths, the NGT has required government departments to come forward and present data and information. This has led to more information and some transparency for assessing environmental damage. But government departments, reluctant to give the NGT information and data, regularly engage in stalling and delay tactics. The Tribunal's conversations with stakeholders allowed for oral documentation of ground conditions. The chairperson's push was also requiring the representatives of the concerned agencies to compare notes on wastewater drains—on their quality and quantity, where they were located, and how much effluent they contained.

Outside of these Tribunal negotiations, the High Courts have had a less interesting involvement with wastewater cases. The exception is the recent rights of nature cases decided in the Uttarakhand High Court. The leading justice was the acting chief justice and used *suo motu* powers to issue a landmark judgment granting rights of nature to the Ganga and Yamuna Rivers and all the tributaries and the glaciers feeding them. These rulings were an earnest attempt to enforce water conservation and contribute to the regulation of wastewater flows but were squarely quashed by the Supreme Court a few months later.[48] In one of these right of nature cases, the justice issued orders to establish sewage treatment plants in cities along the Ganga and mandated action including the closure of polluting industries and institutions such as ashrams (religious institutions) for failure to install their own STPs.[49]

RULES FOR HOUSING SOCIETIES

The rules requiring the installation and operation of STPs in bus depots, housing societies, and other large institutional complexes have followed a different trajectory, originating first in decisions within the High Courts, municipalities, state governments, and the Central Pollution Control Board and then moving into the NGT for further discussion and enforcement across the country. In the state of Karnataka, the government and the Karnataka High Court created rules before they were demanded by the NGT. In 2004 the government of Karnataka set the requirement that STPs had to be installed in housing complexes with more than thirty units. In 2009, the Karnataka State Pollution Control Board modified the rule and required that buildings with more than 20,000 square meters of built space had to set up their own STP. Then in 2015 the Central Pollution Control Board (CPCB) stated that towns across the country did not have adequate sewage collection and treatment and that many municipal authorities had not provided facilities for wastewater treatment. In directions under Section 18(1)(b) of the Water (Prevention and Control of Pollution) Act, 1974, they set the standards for decentralized STPs and required dual plumbing in buildings, to reuse treated wastewater. Specifically the CPCB directed the Karnataka State Pollution Control Board to require that buildings over fifty units have their own STPs and reuse the water for non-potable purposes on their premises through dual plumbing.[50] The Karnataka government issued this CPCB order as a notification to all Karnataka authorities and to the BWSSB (Bengaluru Water Supply and Sewerage Board), the BDA (Bengaluru Development Authority) and the BBMP (Bruhat Bengaluru Mahanagara Palike) so that there would be wide knowledge of the rule.[51] In 2016, the city of Bangalore decreased the minimum number of units in apartment complexes to twenty units.[52] This modification required residential apartments and complexes of more than twenty units, and commercial establishments of certain sizes to construct and operate on-site STPs and meet effluent discharge and reuse standards stipulated by the state pollution control board in compliance with the Central Pollution Control Board.[53] Meanwhile, buildings that had been constructed prior to January 19, 2016, and were connected to the BWSSB underground drainage network were granted exemptions from this rule, unless they violated provisions related to environmental impact assessment. In 2018, the Bangalore Sewerage (Amendment) Regulations established that housing societies with more than twenty units would have to pay fees on top of regular water charges until they set up their STPs.[54]

In the capital of Delhi, the state government was more ambitious on paper, but the results were more like a paper tiger. In 2015 it declared its intent to treat 50 percent of the city's total wastewater by 2022 and reuse 80 percent of it by 2027.[55] In the same year, the NGT took *suo motu* cognisance of pollution in the Yamuna River and indicted the three state governments of the National Capital Region—Delhi, Haryana, and Uttar Pradesh—for poor wastewater management practices. Although sluggish in their response to improve STP performances at centralized and decentralized scales, the NGT forced the authorities in the National Capital Region to zoom in on five-star hotels. In 2017, the NGT ordered eight five-star hotels to reuse their treated wastewater.[56] Meanwhile compliance to rules within housing societies remained weak. In November 2020, the Gurugram Metropolitan Development Authority (GMDA) found that 80 percent of housing societies were not running their STPs and asked the Haryana Pollution Control Board to order them to show their records of STP operations.[57] Shortly thereafter in 2021, the media called out housing societies in Noida, an NCR town in the state of Uttar Pradesh, for failing to install and operate their STPs.[58] This unevenness in performance and compliance will show up in the following chapters, to reveal the complicated landscape in which the NGT attempts to exert its pressure.

2 Inventing Bioreactors

I had been working on problems related to wastewater flows for more than a decade before I started understanding what a bioreactor was. Before that, I saw wastewater treatment plants as sprawling facilities composed of a number of mechanical parts that always seemed to be broken or in a complicated state of disarray or partial connectivity.[1] At that time, I was looking at large centralized facilities that were treating anywhere between three hundred and six hundred million liters or one hundred to four hundred million gallons of wastewater each day. Once I started investigating smaller treatment plants, I could have more detailed conversations with inventors, operators, and monitors. Before turning to the small bioreactors that are the focus of this chapter and the rest of the book, I will explain the three phases of wastewater treatment and then lay out how a large (or centralized) system works.[2] By looking at a larger system first, it will be possible to see how small systems are more uniquely created to suit community and business needs.

Wastewater treatment develops within three phases (primary, secondary, and tertiary) and different grades of usable water can be produced at each phase (see Table 1 in the introduction). In primary treatment, wastewater is digested by microbes in a settling tank or a baffled reactor and the solids that develop as bacteria digest biomass sink to the bottom of the tank. The clear water is removed from the top of the tank and can be used for irrigation and horticulture. Secondary treatment is an add-on to primary treatment and is usually a planted filter bed with gravel and plants to absorb additional nutrients in the wastewater. As this chapter will explain, secondary treatment can also take the form of an add-on called a vortex, which aerates the water after it flows through a baffled reactor. Tertiary treatment would involve primary and secondary treatments and then a final set of filters and disinfection processes to remove remaining solids and kill any lingering bacteria and pathogens. I will provide diagrams and pictures as I move through this chapter, but first I will lay out how a large centralized system works as it includes primary, secondary, and tertiary stages.

In Chennai, for example, the primary treatment stages start with the initial intake chamber and a grit-removing device that removes all the bulk solids in the wastewater, the knotty and sometimes random elements that people throw down the toilet. After the grit-removing devices sideline the big solids and remove them from the water, the wastewater flows to the first clarifier, which is a large tank that holds the wastewater as more solids and particles sink to the bottom of the tank. STP operators remove the sludge that forms at the bottom of this first clarifier and after thickening this sludge with septage from more highly concentrated fecal waste, if it is available from septic tanks, they transfer the sludge to a digester. If the concentration of that septage is high enough to produce methane, they capture the methane in a gas-holding tank and then produce some energy to run the process. Back in the clarifier, the remaining water goes to an aeration tank and a diffuser in the bottom of that tank supplies compressed air to aerate the water. Here the dissolved oxygen or DO is maintained at a specific level to allow the microbes in the wastewater to grow. The microbes use the organic material in the wastewater as their source of carbohydrate for respiration. The oxygen in the air allows the microbes to respire aerobically. After this, the clear water is passed to a secondary clarifier. This digestion becomes "the very heart of the treatment," as one engineer told me. He explained,

> It is a biological treatment. We are maintaining the microorganism. It takes the organic waste as food in the presence of air. We are continually supplying

air to the sewage and we are developing a lot of aerobic microorganisms. They take as their food the organic particles and whatever is in the micro-level form of solids. They generate new cells also. An enormous volume of microorganisms are produced in this tank. So the pure water combined with the enormous volume of microorganisms goes [from the aeration tank] to the secondary clarifying tank. In this secondary clarifying tank, the microorganisms slowly settle down in four hours. Then the pure water comes out of the secondary clarifier and goes to chlorination. In that, we kill all the pathogenic microorganisms. Contact time for chlorine is thirty minutes, but the residue in the released water doesn't affect the aquatic system."[3]

Through this treatment process, the microbes that numbered 20 lakhs (2,000,000) per 100 ml, or milliliters (as mpn, or most probable number), are reduced to 2 lakhs (200,000) in the clarifying process; this means that 80 percent of microbial communities are lodged in and removed with the sludge. Then the remaining microbes are killed by chlorine that is added to the water before it is discharged into the river. In a centralized system, the bioreaction (or digestion) phase occurs during primary and secondary treatment phases. In the secondary phase, the remaining suspended solids or small particles are removed from the water using a second clarifier or by using a trickling filter or planted filter (a tank with gravel that may also have plants growing in the rocky soil to absorb the nutrients in wastewater). In the tertiary phase, sand or carbon filtration, UV light or a reverse osmosis membrane disinfect the water to create the most usable form for human contact uses. It may even be clean enough for drinking purposes. The more the phases of treatment, the more time and energy are involved and the higher the costs go.

This procedure sounds thorough, and it is one of a few different technological processes that are used in centralized treatment plants across India. The other technologies are the sequential batch reactor (SBR) and the Upflow Anaerobic Sludge Blanket (UASB). These other technologies move the wastewater through the system in slightly different ways but still encourage the same processes of microbial digestion in the primary and secondary phases. However, the problem in Chennai and elsewhere is that this intensive process is only undertaken for a small portion of all wastewaters generated in the city.

There are several ways that bioreactions are stimulated during primary and secondary treatment, but usually bacteria are added to the mix

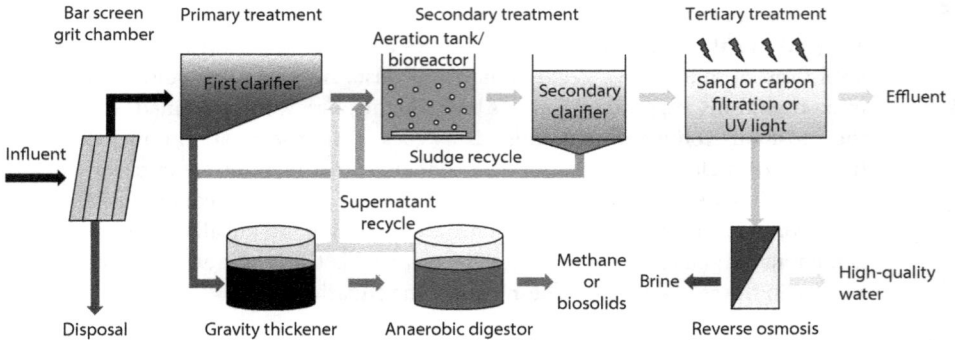

Figure 1. Primary, secondary, and tertiary phases of wastewater treatment. (Created by Shiqianq Zou)

to help the existing bacteria to multiply. A bioreactor is the main container in which microbial multiplication occurs using the infusion of oxygen for aerobic bacteria or containing the anaerobic bacteria in an enclosed tank without the infusion of oxygen. During primary and secondary treatment phases, both kinds of bacteria eat the sewage content in the water and digestion occurs. During digestion, the bacteria reduce the biological oxygen demanding content of the water (referred to as BOD). The biological oxygen demanding content are the solids and particles in wastewaters that suffocate other life and that make wastewater unfit for many safe and healthy uses. The rate at which wastewater requires oxygen is known as the biological oxygen demand (BOD). Aeration and bacteria digestion can remove these solids and particles and then reduce the BOD number per mg. After digestion occurs, the bacteria stick to clumps of waste and they sink to the bottom of the tanks or containers. Then the clearer water is passed through the sand filter and/or an activated carbon filter (made of carbon materials) and then pushed through a container charged with ultraviolet light to kill the remaining bacteria and pathogens. This makes the water suitable for toilet flushing, horticulture or gardening, and washing cars and outdoor equipment. At this point, the water is free of bacteria and pathogens but still has some dissolved materials. To make the water potable, it may be pushed under pressure through a semipermeable membrane to remove salts and any remaining particulate matter. This makes the water good for household and human contact uses and can be used for potable water.[4] After this the water is collected in a storage tank and is ready for release back into the environment or for reuse by humans or industries.

When moving from the big centralized systems and looking at the smaller ones of interest in this book, I have found bioreactors in many sizes and shapes. Small-scale bioreactors must be adapted to limited land spaces and to different kinds of locations, weather conditions, and ecological settings. Many small-scale bioreactors are created underground, below parking garages, walkways, and even busy urban streets. I once accompanied an engineer on a tour of facilities in peri-urban Mumbai and he pointed to a busy road next to a housing society and told me a bioreactor was lying underground, beneath all the vehicular traffic plying along the paved surface. I was astounded! While we are used to thinking of underground sewers, we are less apt to imagine that a bioreactor would be working below the surface.

In another part of Mumbai, I visited a housing complex where the bioreactor was wedged within the thick wall that separated one apartment complex from another. The wall was about ten feet high and five feet wide. Within the four walls of the divider's concrete façade were layers of gravel that acted as the secondary treatment phase. After the wastewater was sprinkled on top of these layers, it trickled down through the gravel. As it trickled down through the gravel layers, the water came into contact with microorganisms that resided in between the layers and those microbes digested the solids, creating bioreactions that led to clumping of the mass inside the layers. The clearer water continued to trickle down through the rock layers. Eventually the water was collected in a storage tank underneath the parking area of the complex. This process, representing primary and secondary treatments, was completely unnoticeable to passersby. Other kinds of bioreactors that look like gravel or plant filters (also called trickling filters) are placed in public parks. There these bioreactors appear as large tanks of gravel with tiny sprinklers and a few large plants residing on top, revealing their true purpose with the faint waft of sewage smell as the wind blows. These decentralized STPs follow the "soil biotechnology method" or SBT, a method its key inventor will describe in more detail later in this chapter. This system is a low-energy method for converting the raw wastewater from sewer lines and open drains into water useful for gardening and for watering roadside medians and vacant land in need of nourishment. The water produced in this process has limited use for human consumption due to lingering bacteria but can be used in places where the public does not congregate. The bacteria and pathogens remaining in the water die off in direct sunlight after two days on the soil surface.

This manner of disinfection is involved with open defecation, a practice now scorned by the Indian government as a shameful act. When defecating in open fields as village residents used to do, the bacteria and pathogens in human fecal waste are killed by sunlight within a few days and the remaining solids fertilize the soil. Today there are greater public health repercussions from the practice of open defecation in heavily populated, urban areas, but it makes sense why it evolved into a cultural practice so long ago. One of the first Hindi phrases I was taught in my Hindi language class in New Delhi was the term *jungle pani*, which named the practice of going out in the field with a small jug of water to defecate and wash afterward. I also remember traveling by train across the agricultural belt in Uttar Pradesh, and while standing at the open door to get some fresh air in the early morning I witnessed a group act of "jungle pani," where farmers sat defecating in a large circle. The practice was indeed sustainable in rural areas, where there was no running water or flush toilets.

BIOREACTOR INVENTORS

During my field visits to see small-scale, decentralized systems, I was able to interview a number of individuals and organizational members who were involved in the early stages of inventing small bioreactors. Some of these individuals are still working in the nongovernmental agencies that consult with communities to find designs and solutions and help them construct decentralized systems. Their fascinating stories lay out the experimentation process, the trials and errors, and the challenges involved in inventing methods within the gaps and beyond the reach of centralized wastewater treatment systems.

One of the participants in the early movement to create decentralized projects was the first science director for the Auroville Centre for Scientific Research. Located in the community of Auroville in the state of Tamil Nadu, the Centre is an international voluntary organization that creates and runs projects in the fields of renewable energy systems (e.g., wind, solar, and biomass), appropriate architecture and technology, wastewater recycling and sanitation, and training. When I met with the director in 2017, he explained their inventive process. During this discussion, he referred to decentralized systems as "DEWATS," which meant small, on-site systems that have minimal energy and maintenance requirements. The use of plants and nature's processes was also fundamental to the notion. He explained:

> CDD [the Consortium for DEWATS Dissemination Society] was conceived here outside under the neem trees and it took a few years to get off the ground after 1995. I was a sociologist and got sidetracked in applied technology. In 1981, I was bothered by the fact that Auroville had no sewage lines and every individual house took care of its own system, basically septic tanks. But after a while the soak pits got clogged on a regular basis. After six months, the septage would not penetrate in the soil and every time you have to clean it out and things like that. So I thought that there must be a way to have the water purified in an easy way and reusable. And that was the start of my long journey in water, in trying to find something that worked which was affordable and which could be upgraded, and which could be duplicated. A lot of experiments from '81 to '95 were mostly failures. We tried several systems with professors who came by and said we have tried this in five-star hotels in Africa and it works. We tried that out in open lagoon systems. Everything. Everything that was possible we did and tried. And usually it failed. It never gave the result. We tried and they failed. Either it smelled or it didn't produce the results. Then we made the compact in CSR with BORDA as partners in an EU project. Natural wastewater system. At the time, they called it low-maintenance, but we changed it to "DEWATS systems." But then we saw that "decentralized" didn't cover it, so we called them "natural systems." If I look back on what we have been doing we have been developing a technology. We are fortunate we have Auroville where we can test out, monitor, and fine-tune the systems at the small scale. Sometime in the mid-2000s, there was increasing interest from outside to implement those systems. We did some and we found that it needed improvement, and those improvements came really gradually. We moved from septic tanks to settlers and then added a buffer tank. And the last one we used to remove the smell, which CDD is using, is the planted filter. It originated from Germany, France, and UK. It was a device that worked extremely well in a temperate climate but did not do well in a tropical climate. There is a lot of activity in the planted filter, cleaning is expensive, and the space needed was not attractive. No promoter wants to invest in a planted filter because of space. There are different names [for the planted filter]: constructed wetlands, root filter . . . different names. But what we

usually call planted filter. After that we had several failures. I knew it was not the device that would carry us forward.

We had a lot of space, but when it came down to new settlements in an urban environment, no way. Like so many times it happens when the creativity in Auroville works together. It is also a long road—the vortex. When I saw it first, I thought this is the solution. It was a long road to make it into a proper device to replace the planted filter. We brought in innovations. It was used to enhance drinking water. It was very tiny. We made it applicable. And then in 2008 to 2011, we actually concentrated on purely research on the vortex, three solid years trying everything. Making it, testing it, and we came to the result that it could replace the planted filter with a minimum, tiny footprint. But it needed a pump. At the beginning, one idea was to move away from the pump and use as little energy as possible. But that is not possible. You have to use pumps in the system. Now we have 100–150 plants, in Auroville and outside, more and more. I view what we are doing as still putting the pieces together to develop a full-fledged technology. We obviously don't have the commercial strength and manpower to go out and say this is the system. We go at a slow pace and every system is custom designed and we have to do a lot of hand holding with those who implement them. The results are getting better and better. We have not reached full maturity, but we are getting closer and closer. And it has its limitations.

I then asked him about the possibility of reusing the water treated in this system. He explained that the first phase involves using this water for irrigation, but they are pushing it a bit further by building toward toilet flushing. He added, "That is the furthest we will go. Not for drinking water. The limit is toilet flushing."

The signature technology of the Centre for Scientific Research is the vortex. It is a device to promote aeration of the water after it passes through an anaerobic baffled reactor where solids settle and anaerobic bacterial growth is generated. After bacteria multiply in the baffled reactor and eat the solid matter, the clumps settle in the chambers of the reactor. The clearer water flows out of the reactor and it is then pumped to the vortex. The vortex is a large, transparent acrylic tube that sits upright on a cement slab at the edge of a lagoon. It is covered on the top with a small vent hole on the side for air. It has a funnel and a diaphragm at the bottom and a pump inlet for the wastewater from the baffled reactor. A vortex, like a whirlpool or a tornado or a waterspout, is created inside the tube and this enables the water to absorb oxygen. The vortex effect increases the

Figure 2. The vortex demonstration setup in Auroville. (Photo by Kelly D. Alley)

exposure of the water to air, giving aerobic bacteria the chance to digest the remaining particulate matter. After the water is funneled up through the tube, it is expelled into a small lagoon. In some cases, the same water is circulated in the lagoon and then pumped up again through the vortex, creating several rounds of oxidation before it is discharged into the environment or captured for reuse.

This group at Auroville works with a few companies that are interested in the low energy treatment process and the easy maintenance procedures. As the director put it,

It is only pumps you have to maintain. Nothing else. There are no mechanical systems. Only the pumps. We have tried a few plants with entrepreneurs
and the difficulty with those organizations, people, or factories is that they
are not very good at maintenance, and they have the habit of doing the
shortcuts. You have to maintain with a schedule. We found that the bigger
companies can do that, and they have that system in place. It is still believed
that if you go to the appropriate tech you don't need to do that and it works
by itself. That is a strong belief in India. You install and it runs by itself. This
is a fallacy. Every system needs to be looked after.

The director added that in Auroville, individual houses take care of their
own wastewater but they have failed to treat sewage at the scale of the
household. He said, "It [the STP] needs maintenance and people don't
take care. Some do but the majority do not. Desludging needs to be done.
You have to maintain the pump and see that there is a flow in the system. Not much but it needs a presence. So the individual approach is not
working, and it is costly. We are trying a much wider system outside." He
continued to talk about the technology of the gravel filter, explaining that
the gravel filter works better in Delhi where the bacterial growth takes a
pause in the winter. A pause is needed. In the South, he said, "We don't
have that [a pause]. It is continuously breathing out in the planted filter.
After six to twelve months, the plants have problems. You have that less
when there is a spell of cold weather. So maybe it works in Delhi. It is good
to see it after a couple of years."[5]

The Centre for Scientific Research at Auroville operates as a legal trust.
In the beginning, they received government grants but these dried up
quickly. Then they started doing construction and consulting to bring in
revenue and that has been going on for thirty-five years. They sell prefabricated baffled reactors and vortexes and that revenue helps to sustain
them and allows them to do new projects. When the radius of the vortex is
not too big, they can load the system on a lorry and transport it to the site.
When the system is bigger, they provide consultancy to builders on-site
and follow up. They give the designs and follow up and provide a schedule
for monitoring. The owner does the testing. They have done setups for two
eye hospitals in Tamil Nadu and these systems use the treated water for
gardening. They do not provide their systems for industrial wastewater.
They cannot treat water to a level required for AC cooling. He went on,

"Milk is too difficult. Fruit processing (jams, etc.) is not a problem. Not paint or chemicals. We have a niche of wastewater that we can safely treat. Single work is more costly than in prefab."[6]

Finally, he commented on BOT (the build-operate-transfer) process that consultants and companies follow when making STPs for housing societies. The failure after transferring a project from the consultant or company to the housing society or association is enormous because the society or association does not have the expertise, the system needs replacements, or the community needs a full-time engineer. Therefore, in many cases, the community stops running the STP or the plant functions improperly. As he put it,

> They have to reach BOD of 10 [in the outlet effluent]. It is really not easy. The PCB [Pollution Control Board] is monitoring and getting rich. They are giving the papers [for a recommendation to operate]. You can achieve a BOD of 10 if it is really optimal. Working with earthworms and planted filters needs maintenance. You cannot leave those plants alone. For us, the main thing is the pumps, and you really have to check the pumps because things go wrong. It is a necessity in a high-tech conventional wastewater treatment plant, but it is even true to a lesser degree in the alternative wastewater treatment plant. You have to take care of them. This is the weak point.

In the following chapters, we will explore the variabilities in the competencies of housing societies. Chapter 3 will show the difficulties that housing societies face when their associations are handed over the operation and maintenance of the STP from the builder while chapters 4 and 5 will show how communities work with consultants to upgrade their facilities or hire a manager for operation and maintenance.

SOIL BIOTECHNOLOGY: LIFE HAS BEEN THROWN OUT OF WATER

The second inventor I introduce is a chemical engineering professor at IIT-Bombay (Mumbai). When I met him for a long interview, he began by making the argument that we should look at this problem of water pollution in the context of the global carbon cycle. Of all the cycles in the environment, carbon is the best understood and studied. He explained:

The energy required to live in water is very large compared to energy required to live on land. Because of the energy it takes and the thermodynamics of life in water compared to life on land, evolution has thrown life out of water. If you look at sanitary engineering over the last 150 years, all the waste treatment is in the aquatic environment. All the technologies have been concerned with throwing waste into water and treating water. But what the carbon cycle is telling is that if you do the same thing in the terrestrial environment, you do not need energy. But if you do it in water, you need a lot of energy. [The] second more important thing is that we are talking about treatment in the aquatic environment. But if you look at water in the world you will not find dead organic matter floating in water. Generally, death is very uncommon in water. There is no death in water, but death is very common on land. The saprophytes which are required for managing death and processing death are abundant in soil. Oxygen is abundant in soil whereas it is not abundant in water. So, you have to use a lot of energy to put oxygen in water. So, there are two fundamental aspects of the carbon cycle that have been ignored in sanitary engineering over the last 150 years. This is SBT. This is the philosophical inspiration for our work over the last 25 years. What we do is fundamentally different. We are talking about an ecosystem which is complete in itself, which uses the power of soils, soil's biology, soil's oxygen availability to treat waste. This is where we are fundamentally different from DEWATS or SBR [the technology called sequential batch reactor]. Both are aquatic technologies.[7]

He continued to do the math on carbon. If you look at the amount of fossil carbon emissions from fossil fuel burning, the total amount is now five to six billion tons of carbon per year. In addition to this flux into the atmosphere, fifty to seventy-five billion tons of carbon are released from the oxidation of organic carbon in soils annually. So if you can make better use of the organic carbon [in the soil] you can reduce climate change. You allow the respiration to occur in human-engineered energy and this energy can be put to use. He emphasized,

If we can live within the carbon cycle by recognizing the power and synergy in it, we will have a better life. Land has abundant oxygen and organic material to produce food. . . . We are supplying oxygen naturally rather than in water mechanically. You need to supply 5,000 tons of oxygen per day for Soil BioTechnology [his team's technology innovation]. Whatever is the quantity of oxygen you need to supply, we can design the media so the oxygen is supplied naturally. Plus there are global warming issues involved with mechanical approaches. This oxygen supply is the most expensive part of the treatment of waste. But we supply this naturally.

He explained that the treated effluent from the soil biotechnology method (hereafter SBT) is rich in oxygen and can go into a pond to feed fish. He lamented that others try to destroy the organics with oxygen in mechanical processes, whereas they try to make use of the energy of oxidation.[8] This method, he argued, is "fundamentally different." Other ways try to destroy the organics mechanically to get clear water. By contrast they are trying to create a resource that can be put to use. He continued,

> Carbon dioxide reaction is picked up by rocks. No technology of waste treatment recognizes the importance of this reaction. . . . As a result, they are not able to integrate this reaction into the process. We do. The pH is near neutral and CO_2 that is generated converts into sand or clay. So we use photosynthesis in this reaction. All these reactions are biologically mediated. To add sustainability into the process, we integrate photosynthesis into the process by making it look like a garden. We don't have to intervene anthropogenically and create an artificial setting. This way the sustainability that is crucial to the viability of the system is ensured. In SBT you see only a garden. It adds aesthetics as well. We can also remove nitrogen doing these two reactions. . . . So we can remove nitrogen and carbon. In the anaerobic technology like DEWATS, the carbon is removed as CO_2 in the absence of oxygen. This reaction produces methane and hydrogen sulfide. We try to remove hydrogen sulfide using a chemical process so that we can use the methane in a burner. In SBT you can also have an anaerobic process. You can have anaerobic, anoxic, or aerobic. But in aerobic you don't produce greenhouse gases, so that is our preference. Soil formation takes ten thousand years in the natural environment but because of the way it is engineered, we can produce it in a few months. The rate of a chemical reaction depends on the particle size. A small particle 30–60 mm in size gives good reactions. So we can produce soil in economic time scale not geological time scale. We can mimic the soil's biochemistry.

The components for their SBT reactors are sourced from materials available in the area where the project is located. In the Ganga basin, they had to change the process to what was available there. They used broken bricks. In other places, they have used weathered rock. They adapt their process and the implementation of the biology to what is locally available. They have a bioreactor and a tank with two pumps, but as he noted, "you might not see the pumps." The bioreactor is three to four feet deep and filled with media, broken bricks, or weathered rocks. The residence time for the wastewater is six to eight hours. The raw water tank ensures that the solids settle. They use flocculants to settle the solids, and as they settle, they also dissolve. So when they harvest the solids, there are no organics. He explained, "We can pump it off once in a year. It is a serious problem in many places. In DEWATS, the reaction rates are low and the organics are very high." In their projects in housing societies, the silt or inorganics going into the water is minimal. The rough cost to install a 100 kld (kiloliters per day) SBT plant is 35 lakh rupees (3,500,000), which as a capex (or installation) cost includes setting up the plant and running it for six months.

During my conversation with him, I described another project I had seen at a nearby hospital. Constructed by a different consultancy, that project used a baffled reactor and a planted filter. I explained that the wastage after coming out of the baffled reactor was congealing on the top of the planted filter. In other words, the water coming out of the baffled reactor still had too much solid matter in it. It was supposed to run through the gravel in the planted filter horizontally but instead, when entering the gravel bed, the sludgy water was just accumulating on the top layer of the bed. This was happening because the baffled reactor was filling up with sludge and not digesting the matter fully. It was too thick to then run through the gravel filter. When I visited that project, I was told by the management company that they had asked the hospital to pay to desludge the baffled reactor (meaning, to remove the excess solids that had over-accumulated in the baffled chambers of the reactor). After hearing this, he replied: "They won't remove. Actually, in our experience with wastewater management, the client has given the problem to you. Please don't come back to me [they will imply]. I have given you money so don't come back to me." I will return to the problems involved with maintaining machines in the following chapters.

CONSORTIUM FOR DEWATS DISSEMINATION SOCIETY (CDD)

A year later, I was able to visit the main office of the CDD, the Consortium for DEWATS Dissemination Society, in Bangalore.[9] I went with a team of students and the postdoctoral researcher to collect information about their operations and for their recommendations on housing societies with operational STPs that we could visit in Bangalore. During that trip, we also visited the offices of former members of the CDD and asked them about the early stages of the CDD. The objective of the founding members of CDD was to display a technology that could meet a part of the sewage treatment needs of the country. As one founding member explained, "It was never meant to be a full treatment. We were never placed as a competent STP. Any reduction was good. It was only meant to reduce pollution, not to complete or comply with the law. We were creating awareness."[10]

The Bremen Overseas Research and Development Association or BORDA is a network of sanitation professionals and agencies based in Germany. They came to India to disseminate their systems in the 1980s. As the director of the Center for Scientific Research in Auroville explained, BORDA members met with leaders under the neem tree in Auroville to brainstorm their ideas. Eventually a number of NGOs were mobilized and formed the CDD in India. The Consortium for DEWATS Dissemination Society eventually included twenty-five NGOs interested in promoting decentralized systems of wastewater management. They used the term *DEWATS* to refer to a range of systems and bioreactors that do not use a lot of energy and chemicals. Starting in 2002, they began developing technologies, and then moved into design and project management. They registered the organization in 2005.

Today, the CDD operates from a building owned by the Rajiv Gandhi Rural Housing Corporation Ltd. (RGRHCL), a government organization working in the field of housing. With their support and the continuing support of BORDA, the CDD developed the Center for Sanitation Information where they carry out training and capacity building programs. They often work with the RGRHCL in the construction of apartment complexes for the urban poor. They are also part of a team of nongovernmental organizations that implemented and now operates and monitors the Devanhalli Fecal Sludge Management plant, the first FSM facility in India. In their main office, they have training facilities, labs, a factory making prefab units, and an exhibit on sanitation. They conduct orientation programs for plant operators. As a key organization working in the field of government drinking water and sanitation, they train government employees and conduct site visits for projects in housing societies.

Figure 3. An exhibition of an anaerobic baffled reactor filled with rock media in the CDD training center. (Photo by Kelly D. Alley)

Of the one hundred people working in their main office, 50 percent are engineers.

When focusing on community-based small and medium-sized enterprises, they usually work with a community organization, which eventually takes the project over for long-term operation and maintenance. They help them to collect data, design the system, implement it, and monitor its functions. Once their project is completed, they train a community member to operate it. At times there are changes, and they provide an explanation of how to modify the operations. They call them to their office and conduct training. They also work with the RGRHCL to set up projects in the low-income settlements where the RGRHCL is the sponsor. A few of these projects will be introduced in following chapters.

They describe their technology as a natural biological system, but they are flexible on the design. They may select an improved septic tank with up to five chambers so that the wastewater flows up and down through

Figure 4. CDD model STP in a residential community. (Photo by Kelly D. Alley)

the chambers. They also use a planted gravel filter or wetland system with plants for secondary treatment. However, the land requirement for wetland filtration is greater. They construct plants with a capacity range of five hundred liters to one hundred kiloliters per day. They will design a component for reuse and for the most part will suggest landscaping as their primary reuse option. While some housing society members also want a reuse option for toilet flushing, the CDD member explained that it depends on climate requirements and land. They can combine wetlands with a tertiary treatment such as chlorination to kill off more bacteria. If they design a unit to include an aeration device and sand and carbon filtration, then more power is needed to run the system.

The CDD member explained the challenges:

> So what happens is that people are fascinated with the idea of a low energy requirement. People ask for low energy, but they do not know that there is a land requirement. Then they say it is so expensive because they don't realize the space requirement. That is the biggest challenge. We need space for

retention time. They don't have land, but they want a low-energy system. So, we convince them. They say Madam, we have a parking lot, so can you help us do it? Their structure is already there, so you have to work with it. That is more from a design perspective. In operation, people know it is low requirement, but they think there is no requirement to remove the sludge. They think they do not need to remove the sludge even after five years. Also, with maintenance there is a challenge. In an anaerobic system, there is gas. In some systems we have a vent pipe, and they need a bit of maintenance and they do not do it.[11]

She elaborated that there were safety issues involved because the gas builds up in an anaerobic system. When the pipes get choked, a person needs to visually inspect them. "We tell them not to enter the pipe and to use gloves. We explain how to use the biogas and how to remove sludge. We explain that the sludge requires drying in sunlight and show them how to reuse it. An anaerobic system needs two to three months to stabilize. We do analysis on that and if we are doing O and M [operation and maintenance] then we monitor every month." She continued:

Requirements for water now influence the client as they have a better understanding of their requirement. They know their wastewater situation and have water needs. They have studied and then come with their needs. [They say], "Madam, we know your tech will not treat up to the mark, but we are interested." We show them the systems and they are convinced. We discuss the obstacles in terms of site location and land. We discuss what can be optimized. We do R&D with them. The audiences are more aware of wastewater treatment now. The peri-urban areas are growing fast, and they know they don't have water and they have to spend a lot with tanker water. And developers do lots of good landscaping and there is toilet flushing. Since there is a demand for water, wastewater can be a source. It is the policy and the water scarcity in urban areas has created the push. If there is a group responsible for the system, they operate well.

In chapters 4 and 5, I will present the engagements of members of Resident Welfare Associations (RWAs) and their management staff concerning the functioning of their STPs. We will see that modifications and optimizations are regularly made to reduce odor, particulate matter, and other imperfections in the treated water. Some of their statements will also demonstrate what this CDD representative is saying about the general desire for low-energy, low-maintenance systems. In many cases, low maintenance is impossible, and communities must grapple with how to allocate time and energy to project sustainability.

ECOPARADIGM

The last inventor that I will introduce is a founding member of a consultancy named Ecoparadigm. He was involved in the first group of NGOs brainstorming on decentralized treatment technologies under the neem tree in Auroville. The founding director gave our research team a history of his involvement in projects over the last fifteen years and communicated his perspective on the way responsibilities have been distributed over time to communities and interest groups. He explained that, at first, there were many interested partners and NGOs to create awareness about DEWATS, and to set up projects and demonstrate them. In 2005, he designed the MDCO complex as an independent consultancy job. After that, he worked at CDD and they developed a system in the Manipal hospital. It was a public-private partnership, a "PPP" project. He recounted:

> In 2007, we designed the Manipur hospital with the most stringent standards, less than 10 BOD [mg per liter]. Now it is relaxed to 20 BOD. Manipur and ATREE [another project] were designed for 10. We were reusing before it was customary to reuse. We used our own tech. . . . The hospital was monitored by the High Court of Karnataka, which imposed a stringent standard. This is related to a 2007–2008 High Court case. I don't know what provoked that. During th[at] time, the law was that all institutes could discharge into sewer lines if their BOD was less than 350. They had to reduce it to 350 before discharging into the sewer that went to the centralized STP. They had to treat it only if there was no [centralized] STP at the end of the pipe. They had STPs then, so they were able to discharge. But the High Court intervened and said this is not OK for a hospital. So many hospitals were issued closure notices after this project was successful. The Court was of the view that since it worked in Manipal then others have to do [the same]. Lots of hospitals in 2009–2010 installed STPs. After this, the Martha's charitable hospital, was commissioned in 2009–2010 for a 350 kl plant. Then they wanted to add up to 850 kld. We upgraded it from 350 to 850 at a fraction of the cost added over time. The quality is very good in the treated water. They spent 5–6 lakh rupees (500,000–600,000) to operate it for eight to ten years. It is extremely negligible.[12]

The client makes the tender and selects the contractor and then Eco-paradigm members train the contractor to make the design and supervise them during construction and during the maintenance work of desludg-ing. The Ecoparadigm group claims that their baffled reactor does not need an attendant or specialized contractors. They avoid the planted filter because many clients do not have land.

He explained that they encounter two types of situations: when the builder has to install an STP by law at the time of construction or when there are site restrictions. Before 2015 when there was no real enforce-ment of the STP mandate for housing societies, the builder had to create a simple structure to minimally enact compliance. He admitted, "In some rocky area, the builder had to make an STP and discharge the treated ef-fluent on the road." When the housing community then later decides that they want to use the treated water for flushing toilets, they realize that they have no way to get the water to the flush toilets. In most cases, the builders had installed the STP and avoided installing dual plumbing, as they were not abiding by the 2006 law. But now, he said, clients are coming to Eco-paradigm because the builder has not constructed the STP properly and the PCB is slapping conditions on the owners within the housing com-plexes who make up the RWAs (Resident Welfare Associations). At this point their problem is more complicated because the builder never allo-cated ample space for a proper STP. So Ecoparadigm members have to work within the space constraints. With the parameters already set, he explained, they have to use the parking lot. He went on to elaborate their design activities:

> We don't have regular shapes like a rectangle or square. We design in all shapes. We find complicated wastewaters in hospitals—antibiotics, deter-gents, etc. Flow zones are not uniform. Before [the] law became strong in housing, the builder would make a small settling tank. Then later it would overflow. So, clients contact us at that time. They say they want an STP but don't have more land. The DTS has to be within the footprint of the [origi-nal] septic tank. We have done three [STPs] within XX Housing Society to design for 10 BOD. It is similar to the CDD model with a baffled reactor filled with rocks. No blowers. We use rocks. We don't want to be constrained by anything. We can use bottlecaps, stone, grit, cinder [inside the baffled re-actor], whatever is local. The system is anaerobic. It is like a settling tank with two to three compartments. You need specialized bacteria in each tank. If you want a system to work, it needs to work to my efficiency. There are

specialized bacteria. There is a consortium of bacteria. Not a single bacteria. Yes, that is our patent [the specialized bacteria]. Some [reactors] need one kind of bacteria, another a different flow rate. Each bacteri[um] has a particular favorite condition. You have to match [it] with the situation. These are naturally occurring. We have concentrated on that. You have to introduce it in the beginning. Afterwards it is straightforward. There is a way to resume the bacteria. We have specific bacteria. That is what we make. We select specific bacteria for hospitals.

Now in the discussion the microbial dimensions of a treatment plant come into full view. When we were listening to the explanation, we were rather skeptical as laypersons. What exactly are the bacteria he is referring to? We wondered about it and talked about it as a team later. Are the bacteria really that specialized and how can we tell what is going on with them? In the following cases, other STP operators will mention bacteria so that the critical role of microbes can be at least thought about if not really seen.

Finally turning to the problem of odor, he added, "These days it is complicated. It is difficult because they will complain. We just put [in the STP] a media filter. It is a placebo thing because what else can you do?" He said they could install carbon and sand filters, but the clients rarely have enough land. The problem of land is a critical one. Allocating valuable land for a nonfunctioning or low-functioning STP is not a desirable option for many. If reuses can be realized, then the land allocation has a bit more traction. We will see a variety of responses to land constraints in the following case studies.

This chapter has introduced a few of the key innovators in the field of technology design, construction, and consultancy. These actors know each other, and many have worked together during their careers. They form a network and together they build the working knowledge of what has worked and what has failed. They tie up with government agencies or work independently. Their practices are based on scientific experiments promoting microbial activity within machines, adjusting machines to remove solids and then killing off those bacteria, and building awareness and operation and maintenance capacity among communities. These innovators must adjust their bioreactors to specific site conditions and use materials sourced from these locations. They understand state and central regulations, but they make no mention of any centralized policies dictating the methods, shapes and sizes of their bioreactors.

3   Double Burdens

As inventors went along experimenting with technologies, communities started to work with scientific, nongovernmental, and consultancy agencies to adopt them. At first, most projects were targeted at low-income groups. I say "targeted at" because many of these communities did not request the projects but were introduced to them as part of a plan created by the builder, or part of a resettlement program created by a government agency such as the RGRHCL, or part of an experimental model such as those created by CDD. Given the uneven water services throughout urban centers, projects in wastewater treatment and reuse were adding another burden on the existing burden of procuring water. In this chapter, I discuss three communities that represent the variety of challenges facing low-income communities as they deal with the double burdens of water and wastewater management. These burdens end up sinking their chances of developing reuse activities to alleviate some of their water scarcity. The three communities vary in size, leadership power, and water budgeting, and their situations involve social contracts with state and nongovernmental agencies in the provisioning of public goods. The descriptions will show the many tangled reasons for their failing STPs.

Disparities in water and sewerage services across the country are based on geography, location, and income.[1] Disparities are also created by the absence of centralized water and wastewater grids in many peri-urban areas. In off-grid locations, residents must find their own water sources and create their own sewerage facilities by updating or augmenting the facilities provided by builders, contractors, NGOs, and government agencies.[2] The other disparities arise from differing abilities to arrange for water and sewer services, which are de jure the municipality's responsibilities but de facto become their own. It is in these service gaps that communities are forced to organize to secure their own water supplies since no individual can do more than provide drinking water on his or her own. Wastewater services are much harder for low-income communities to self-serve when they are already battling water scarcity. Before exploring the problems, a brief overview of peri-urban water struggles is required.

WATER IN THE PERI-URBAN CONTEXT

Since the 1990s, a shift in the urban population has been underway; by 2015 the urban population reached 420 million or 33 percent of the country's total population. The megacities of Mumbai, Delhi, Chennai, Bangalore, and Kolkata have been expanding outward, proliferating peri-urban zones around the urban cores. Across the country, towns have been mushrooming to form the second tier of urban centers. Between 2001 and 2015, the number of cities with populations of more than a million increased from thirty-five to fifty-three. IT parks built in peri-urban areas were also grabbing more of the share of public water and sewer goods, adding another strain for municipalities delivering essential services.[3]

In the peri-urban areas of Delhi, Mumbai, Chennai, Kolkata, and Bangalore, there are limited piped water services, and few sewer lines to functioning sewage treatment plants. At least 2.2 million residents of Bangalore's population of 12.3 million do not have access to piped water.[4] In south Delhi, high-end apartment buildings are built in areas where the courts and government have banned the pumping of groundwater. Groundwater has helped burgeoning communities survive the rapid growth, but levels are depleting in most urban areas.[5] The quality of groundwater is poor and may not be suitable for drinking or bathing. In some locations where groundwater levels are critically low, the NGT and the Central Ground Water Authority have issued bans on extraction. In situations where groundwater is depleted or banned, residents must purchase water from private companies that supply it through bottles, cans, and tanker trucks. The prices charged for "ATM water" or "tanker water" are generally higher per kiloliter than the prices charged by the government for piped water. While ATM water is usually of potable water quality and sold in local reverse osmosis centers or dispensaries, private tanker water is of questionable quality, as there are no monitoring agencies to ensure that it is adequate for household consumption. ATM water has some quality assurance from the bottling companies, but it is more expensive than tanker water. During the summer months, residents scramble to purchase private water supplies and the tension to pay for them causes increased stress. These conditions put water availability at the center of motivations to participate in community schemes to produce reusable water close to home.

"THEY CAME AND WENT AND NOBODY IS HERE"

The urban core of Bangalore, covering 800 sq km and 80 percent of the metro area, is sewered.[6] Within the sewered grid, there are twenty-five centralized STPs operated by the Bangalore Water Supply and Sewerage Board (BWSSB) and the reported total capacity is 800 mld.[7] But there are not enough pipelines to bring that much sewage to the plants every day, so the real treatment total is around 500 mld. This means the centralized STPs are "underloaded."[8] According to the CDD, Bangalore has 4,000 decentralized treatment systems in neighborhoods, housing complexes, hospitals, universities, and other large institutions. Along with the CDD and Ecoparadigm, there are additional service providers in Bangalore, Mumbai, Chennai, and Delhi working in decentralized wastewater management.

In Bangalore, the research team visited a peri-urban housing community on the northern corner of the metropolis. We borrowed our guide from the CDD to help explain the infrastructure to us. On our arrival at the site and before speaking to the residents, we asked our guide to show us the infrastructure parts and connections. Since everything was underground or hidden by cement manholes and slabs, we needed someone to explain the underground circuit. First, we walked over to a large cement slab about the size of a small parking lot, lying adjacent to the apartment blocks. We were looking at the roof of the baffled reactor just below the soil surface. Our guide explained that there were three chambers underground.

In the following conversation, we were trying to figure out what was happening underground. Since these underground realities are generally murky, I use excerpts from our conversations to convey the general process for initially investigating the infrastructure. It was important in each location to understand the infrastructure grid before starting to talk with local residents.

CDD REP: This is the distribution chamber, the ABR [the anaerobic baffled reactor] outlet. From here it goes here. [He points to the inlet of the ABR tank. We are mumbling about the smell and that it looks dirty]. Meaning it is stagnant now.

GRA #1: Why?

CDD REP: It doesn't go out [the sewage does not exit the ABR tank].

KA: From here it all goes to the ABR. [Pointing to the sewer line that attaches to the ABR]. But this is nothing . . .

GRA #1: Why? It is not working right now?

CDD REP: Yes. Somewhere it is choked right now. It doesn't go up to the drain [the wastewater does not get filtered through the ABR and exit into the nearby drain]. They haven't talked to the next-door property about it.

GRA #1: So since three years?

CDD REP: From the last two years.

GRA #1: From the last two years it is working?

CDD REP: From the first year it worked and then not after that. After that, there was some problem [garbar] and then nothing. All blocked mostly . . .

GRA #1: So CDD doesn't take care of that?

CDD REP: Yes, they will take care. We [the CDD] told them [the government agency in charge of the settlement construction] to please do this and that and they didn't do anything.

GRA #1: CDD set this up? Who paid the money to CDD?

CDD REP: RGRHCL [the Rajiv Gandhi Rural Housing Corporation Limited].

GRA #1:   RGRHCL paid the money. And RGRHCL doesn't inspect this?

CDD REP: And sometimes RGRHCL came, and they said we will do, we will do. But they didn't do anything.

KA: What is the organization?

GRA #1:   It is the RGRHCL. They have paid the CDD. Can you say what the expenditure was to set up this system?

KA: Yes, you can smell it and it is not working.

GRA #1:   Yes, because it is stagnant and not working. It is all choked up.

CDD REP: Yes. They don't provide an outlet [meaning the contractor never built a way for the wastewater to exit the tank after anaerobic treatment].

KA: The inlet [for wastewater] is coming in but not going out.

CDD REP: Not going out. They haven't given the connection over there to let the sewage out [into the nearby open drain].

NM: [Arrives and wants to get caught up on the discussion]. Wait so I am just wondering, it is working or not?

CDD REP: No, stagnant and not working.

KA: He says there is some [wastewater] in there from before. It is coming in but not going out.

NM: How long is it not working?

CDD REP: Some six months, I think.

NM: And still that sewage is coming from there to here.

CDD REP: Yes. And again, it was working. Then it turned blocked.

NM: Blocked. . . . It gets full here. Where?

CDD REP: Over here I am saying. Here is one inlet. It is blocked. Nothing goes in. Here the black water does not go in.

GRA #1:   This is not working. So what happens to their sewer, those who are connected?

CDD REP: After a little while it gets choked.

NM: Wait, tell us. It doesn't work. It gets choked. What happens to the water that the residents use in the house?

CDD REP: I am saying it can change when the line gets choked. After a little while, it goes to another lane. It will get changed to another lane when it gets choked in this one.

NM: How does it go in another lane?

CDD REP: It will go in another sewer lane when it is choked and then like that. When it gets choked, that is, it will be put in another lane.

NM: Oh. So where is the underground sewer line for this water? [We walk to the other side of the ABR.] This is the outlet?

CDD REP: Yes, that's the outlet.

NM: Where does this go?

CDD REP: Dumps into the nala [open drain].

NM: Goes straight to the drain? You haven't put up a phytorid plant [wetland system] or anything?

CDD REP: No nothing.

KA: It is blocked here and oozing out here on the edge because you can smell it.

GRA #1: So when CDD handed it over to the RGRHCL you did not make the outlet? Like you said, the outlet was on a private property. Was it the responsibility of the community to make the outlet formation or CDD?

CDD REP: See this is not our property [meaning the property adjacent to the open drain near the ABR tank].

GRA #1: Yes, I understand. I am trying to figure out how it works.

CDD REP: They have to do. They [the RGRHCL] have to do. We are chasing and they are not responding. CDD also. Means they say we are doing, doing. Something like that. Someone came and went. Nobody is here.

After this tour, we started meeting residents of the housing community. We began with a small focus group and then interacted through the verbal administration of the survey. The questions gathered demographic and cost data for water, electricity, STP installation, and maintenance, and elicited responses to three reuse scenarios produced by the three levels of treatment—primary, secondary, and tertiary. It also gauged willingness to pay for system upgrades using five bid options for each scenario. The survey was also an open-ended instrument that generated a lively discussion on problems and hardships.

After documenting the responses, the research team was able to cobble together the bigger picture. Some residents told us that the wastewater from their toilets and from the sewer lines was backflowing into their bathrooms when the sewer lines got choked. So we asked, "When the water doesn't come out of the bathroom what do you do?" They replied, "We pay for a tanker to come and suck the sewage out of the midway tank—'the chamber'—outside the bathroom." We were taken to see the

Figure 5. A clogged anaerobic baffled reactor in a low-income community. (Photo by Kelly D. Alley)

midway chamber and continued to ask questions. Who cleans this and how many times? One resident explained that every three months he pays someone to come and suck out the sewage with a pump and haul it away in a tanker truck. He then asked, "Are you all from the media? Why are you asking this? Can you solve this problem? Over the last five years no one has maintained this [the infrastructure]."

Continuing with the survey, we learned more. A few residents explained that one of the representatives living there had written to the RGRHCL "at least fifty times" but did not receive a reply. Others chimed in, "There is no drinking water here. In fifteen days they give us water one time." They continued explaining their conditions, "The maintenance is supposed to be under RG [now abbreviated from RGRHCL] but they have done nothing." They procured water from a water ATM installed by the local government and paid six rupees for a twenty-liter can. They also paid Rs. 350–400 for each private water tanker containing 6,000 liters. One tanker was shared by twelve houses, and a few tankers arrived every day. Each household paid around Rs. 1,000–1,500 per month for tanker water supplied by private companies. One resident said that she filled her overhead tank with the tanker water each day and by the end of the day, the tank was empty. Several residents pointed across the street to the aboveground water tank and said, "There is one borewell over there owned by the government [and operated by the RGRHCL] but in fifteen days we get water once from there." The unevenly provisioned government borewell water was free but each family paid Rs. 1,000–1,100 per month for the electricity to pump the water to their individual rooftop tanks. No one was cleaning the nala (the open drain nearby). According to them, the RGRHCL was supposed to do it, but they did nothing. A smell was coming from the clogged sewer lines.

The residents explained that when the decentralized wastewater plant was first built the RGRHCL took a payment of Rs. 75,000 from each resident for the construction. They had to give this deposit before they could move into their houses. During our visit, there was some discussion on whether this amount covered the maintenance of the STP, also called the DEWATS facility. "They took extra funds from us [Rs. 75,000] to make the DEWATS and for other functions." Then one younger member of the community chimed in to summarize the whole state of affairs:

> The manhole shows there is just one inch left before it overflows. Our sewer section is already blocked. It will be 100 percent blocked at any moment. At any point, it will overflow. One of our neighbors is leaving the house because of the smell. No one is maintaining. If it is RG or the village panchayat, no one is maintaining. We are suffering. Once it comes out, we will not be able to stay here. These houses are living beside the DEWATS, and they get the smell. They cannot sleep on that side of the houses. They get infected by fever. No one has visited since they built this place. We had water tank issues [with the rooftop tanks] but there is no ladder to get up there. We couldn't fix our water tank, and for six months we didn't get any water. Then we got it fixed. No one is taking initiative. They have not built the lines in a way that the sewage goes out. It just gets stuck there. The [borewell] water is salty. It is so hard, salty. You cannot take a bath in that. Can't drink it. Some people don't even get that water. There are kids and elderly people. Every house pays Rs. 250 per day for tankers and up to 1,000 rupees per day for all the water supplies. No one is ready to supply the water for this colony. There is

only one locked borewell that we cannot access. It is controlled by one RG guy.

While we were discussing the problems, the RG guy drove by on his scooter and a resident began yelling at him. A few minutes later, the water started to fill up in the common water collection tank. Our presence as surveyors had incentivized the RG representative to start the borewell pump after what the residents claimed was a two-week dry spell. Meanwhile the smell from the manhole blew past us in the wind.

After the tour and these discussions, we could piece together the bigger picture. We were able to find that in this location a Bangalore-based nongovernmental organization had created a baffled reactor to treat their wastewater. This is a large septic tank with many chambers like the CDD unit diagramed in chapter 2. Unfortunately, the reactor got clogged and so did the sewer lines leading to the reactor. They have not been desludged or unclogged since the first incident, and since then the sewage has backflowed into their homes. Some residents could not flush their toilets without creating a backflow. The community did not have the financial capability to get the baffled reactor cleaned (to get the sewage sucked out), but one resident was able to pay to have the wastewater from his collection chamber sucked out every three months. It appeared that his chamber was at the lowest-lying area and the sewage from the other apartment buildings was collecting there. This resident was carrying the load in terms of paying for the basic cleaning functions of the septic system. The ABR, on the other hand, did not have a functional outlet point due to a dispute with a property owner nearby. The property owner did not want the wastewater from the treatment plant to run into the drain near his house. Given his apparent wealth and power, he was able to block the creation of an outlet for this ABR. So the RGRHCL gave up the task of creating an outlet for the treated water.

All these infrastructure details meant that the ABR for this housing community was just a large septic tank and did not perform any meaningful microbial digestion and treatment. However, in its clogged state, it did not even function as a septic tank and the resident's low-lying chamber became the catchment for this mangled sewershed. While residents were caught in the constant struggle to get water, governmental and nongovernmental agencies were failing to resolve the infrastructure problems.

One resident was using his own funds to drain the clogged lines when they backed up.

After the interviews, surveys and discussions, our team visited the Devanahalli fecal sludge management plant nearby to get more information about who was responsible for what at the previous site.[9] There we talked to a representative of the CDD who was on staff at the plant. We learned that the Karnakata Habitat Center, the original contractor for the RGRHCL, had made the sewer lines and the baffled reactor. While the CDD provided the design, the CDD representative claimed that the contractor did it incompletely. They were supposed to lay an outlet line and connect it to the raja kaluve, the large open drain nearby. She explained, "Supervision was very bad. We have mentioned that they were not following it [the design specs]. RG will call us, and we will give them the recommendation." I followed up, "Has the sludge ever been removed from the ABR by anyone?" "No," she replied. "They don't want to desludge. They are supposed to dump in the drain after treatment. Since they are not doing, it is backflowing." After I described the full scenario to her, she agreed that they needed to desludge. Who would desludge then? Who would be responsible? She replied, "The housing association members have to do it. We have handed it over to the housing association." Apparently, that also meant that the housing society was to be responsible for the maintenance. I asked further about who would know about who is responsible for the maintenance? She replied, "That I don't know." I pressed, "Who would have documents about who was handed it and who is responsible?" She deferred, "The RG people only. Now so many people have been transferred and I don't know the MD now."

At first glance, this case might be solving the problem of freeriding that Olson and others found problematic for success in collective action around essential services.[10] The fact that one community member was taking care of the desludging could support Olson's argument that small communities can carry free riders while also distributing benefits to members. But there were other obstacles to collective action in this community that were tied to understandings of responsibility and agency. The responsibility for maintaining and cleaning (also called desludging) was in dispute. The residents appeared to think it was the responsibility of the government RGRHCL since they paid the initial fee for the STP, while the CDD representative at Devanahalli thought that upkeep was the responsibility of the residents.

The residents were unable to negotiate the tangled arrangements between the agencies that originally built the treatment system.

Financing the maintenance was another problem. Generally, housing complexes are governed by their own housing societies or resident welfare associations (RWAs). In this low-income community, the RWA was nonexistent and monthly fees were not collected for building and infrastructure maintenance work. Each household had paid the upfront amount of Rs. 75,000 for setting up the STP, but nothing more was decided. The residents were caught up in paying for their basic water supplies, which they had to arrange with private tanker companies. They were also straining to pay the electricity costs of pumping the meager amounts of water they received from the RG guy to the overhead tanks. The interviews with residents made it clear that they did not have the wherewithal to manage anything more than depending upon one resident to get his low-lying chamber desludged every three months, a patchwork effort in a tangled system of low government accountability.

Olson has argued that population size may be a determining factor in the success of collective action.[11] Individuals in large groups will lose the incentive to contribute to a public good because the benefit to each individual will end up being quite small. By contrast, small groups may have a better chance at collaborating to provide a public good, since each person would receive a larger benefit of the public good and free riders (those who do not pay into the system) would be carried along by group leaders. However, in this case, the community's inability to sustain the treatment system and consider any reuses from it was directly related to their struggles for water and their financial incapacity to desludge the main baffled reactor. Their small group size was not a benefit because the funds they needed to raise to run the treatment and reuse system surpassed their ability to pay as a group. Some of these features will appear in the next case as well.

"THIS IS A 'COMA PATIENT'
AND OUR LAND IS VALUABLE"

During the same trip to Bangalore, the research team visited a low-income community on the other side of the city that had a stronger housing association. This housing association was part of a statewide Muslim association, the Karnataka State Beedi Workers Multipurpose Cooperative Society. The head office for this state cooperative society was located within the community and its main officers were in charge of ensuring the water supply for residents. The community of thirteen four-unit apartment buildings sat on eleven hectares.

In 2005, the Central Pollution Control Board, wanting to create an experimental decentralized (DEWATS) project, built an STP within this colony. They operated the system and collected data from the operation for a few years and then handed it over to the RGRHCL which operated and maintained it for another three years. After this, the RGRHCL handed it over to the resident housing society, but the community had difficulty managing the project. In 2012, the CDD took it over and used it as a research facility. In the first two years of operation under the CDD, the STP produced gas from methane, but the community members were not using it consistently and the lines got choked. None of the institutional operators were able to take their treatment process to the next stage of producing reuse water.

The colony did not have municipal water but had two borewells from which they pumped up groundwater every day and stored the water in a community water tank.[12] The community was also purchasing ten to twelve tankers of water from a private water supplier every day for Rs. 350 per tanker. Each tanker contained six thousand liters. In total, tankers were supplying sixty to seventy thousand liters every day. This tanker water combined with the borewell water provided about one hundred twenty thousand liters of water every day to the 4,500 persons living in the colony. We took our estimates from the community RWA leadership because there were no meters on their borewells. The government electricity department was providing free power to pump up the water, a deal made with the community through its leadership, to provide a public good. These free water and power deals made by state governments are what Ho has termed socialist, populist social contracts, the informal arrangements that enable governments to accomplish certain tasks.[13] According to Ho, the Indian state operates according to a socialist and populist social contract in which citizens expect a certain amount of free goods from politicians who need to buy their votes. The fact that this community was also a minority community with a strong leadership meant that there was additional solidarity to draw in support, which then benefited the entire group. However, the notion of a populist and socialist contract is too broad for the picture this ethnography is starting to paint, where there are many strategies at play to procure the public goods of water and wastewater services through central and decentralized infrastructures. The previous case showed that the water public good was provided sporadically by the state agency, barely maintaining the social contract, while other government agencies failed to prevent the original infrastructure from dismembering. This is the unevenness of social contracting within the administration of water and sewerage public goods in the country.

In this second case, there was more to learn. Every house in the community was of equal size and each home had an indoor living space of three hundred square feet with an outside courtyard for washing kitchen and laundry items. The water pipe for each home was in the outside courtyard and near each tap there were two large barrels for storing water. Residents were paying Rs. 150 per month in what they called maintenance fees. This payment covered their water supply and other community services such as garbage collection, street lighting, and cleaning. Every four days, each household received twenty minutes of piped water from the outside tap. Each family filled two drums from this piped water, and this constituted their water supply for four days. This water was supplied on a fixed schedule and residents did not complain about the supply or the timing. Through the uniform survey questions, the research team was able to confirm these characteristics, and that each household was getting the same amount of water and using it for non-potable purposes such as washing clothes and the floors of the home. Additionally, each household purchased RO water from a Water ATM installed by the Government of Karnataka. Before that time, they had been using the community borewell and tanker water for their potable needs. Now at the ATM, they were paying 5 rupees for each twenty-liter can. Each house was paying a monthly electricity bill of around Rs. 700–900 for their household needs and did not have charges for pumping groundwater. There were no sewer charges.

When asked about the sewage system, residents were a bit confused. Some perceived that the toilet water was going to the "chamber" (their name for the STP) through an underground sewer line. One group of respondents said, "It goes over there to a 'gas plant' [the CDD plant]. It filters over there."[14] Generally residents did not perceive a sewage problem. They did not know about the possibility of recycling the water and were not interested in the idea. Even though they were living so close to a model DEWATS project, they had no information about its functions. After explaining how recycling of the water could work, a few residents decided that recycling water sounded like a good idea because they experienced water scarcity. However, some residents continued to say that they would not reuse the water even if there was a good system in place. One resident qualified, "If we know the water is coming from the latrine then we do not want to use it." Their answers to our survey questions revealed that they did not have information about reuse and generally felt that the treated water would be harmful to their health. Household uses of recycled water were considered unacceptable. However, they did agree with the idea that treated water could be used for street cleaning, washing cars, and flushing wastewater lines.

As our team continued with the survey questions, residents began to think more about the options, and appeared to warm up to the idea of reuse. We asked if they would be willing to pay an extra amount per month for three grades of treated water. The first payment bid was to produce usable water from primary treatment for horticulture, the second payment bid was for water after secondary treatment that could be used for toilet flushing and horticulture, and the third payment bid was for water from tertiary treatment that could be used for household cleaning, flushing toilets and horticulture (see Table 1 in the introduction). They started answering these questions by rejecting the proposed reuses of treated water. Then one woman said, "OK, Rs. 150 I will pay." Then another chimed in, "Hey hello! Would it be available 24/7?" It appeared that the possibility of continuous water supply was appealing, especially if the water could achieve the grade produced by tertiary treatment. They began to hedge, "If it is good water then we can use it," as they softened up to the hypothetical idea. "Yes, we would reduce use of the good water [tanker and borewell water] if we used the recycled water." However, I realized it was difficult for them to visualize the reuse of this kind of water because they had not seen it or experimented with it. The 'willingness to pay' questions were

stuck in hypothetical thinking.[15] Nevertheless, the conversation was easing them into thinking about reuse in times of freshwater scarcity. They
still needed the details on treatment and safe reuses before considering
anything further.

Turning back to the DEWATS project, one community leader in the colony said to us, "Now this DEWATS plant is 'a coma patient.' Cooking gas
was supplied for free but in the last five years there has been no gas." The
leader continued to complain that the DEWATS was "only for publicity."
At the time of our visit, the community leaders informed us that they had
given notice to the CDD, demanding that the land on which the STP stood
should be returned to them. They claimed that the value of the land was
10 crores (Rs. 100,000,000) and to keep such valuable land for publicity
produced no benefit. "It is *bekaar* [useless]." The leader added, "We gave
them [the CDD] a target of September 25, 2018, to remove the plant. We
do not need to go through the courts, as this is our society's property. . . .
We are not getting anything from this land now. There is no need for this
plant. This land is society property, and we are the governing body. CPCB
and RG gave the plant, but it was our land. They told us we would get gas
and recycled water. But then there was nothing. Just a picture. The CDD
boss changed."

In this second community, several complexities arise. The homeowner's
association members were part of a statewide network, and their leadership was able. Using funds raised within the community, the leaders
could purchase water supplies from private tanker companies and draw
up unpriced water from their unmetered borewells. Their social contract
with the government centered on the power supply, which indirectly reduced their water bills since they were using free groundwater. There
were no wastewater services, and it appeared that residents were sending
their household wastewater to the sewer system and the flows were dis-
appearing somewhere underground, perhaps running into the STP but
probably joining a drain along the edge of the community land. We could
not find that flows were entering or exiting the on-site STP. Since community members were not paying any maintenance fees for the STP, there
was a lack of knowledge about its usability. The CDD was maintaining
the facility as a model plant, but without any connections to community
input and interests. Instead, the STP appeared to residents as a stage on

which the agencies appeared to be meeting local ecological and waste-water management functions. But it was a false system since the STP was not treating much wastewater and not generating any reuse options. The residents showed some interest in reuse options during the survey, but the leadership was more interested in the value of the land, to develop it for other purposes.

As Auerbach has explained, in low-income slums variation in the provisioning of public goods is the result of political organization and the degree to which a community is integrated into larger party networks.[16] Larger settlements are able to attract the attention of party leaders easier than smaller communities can. The first and second cases show that free water or power allowances are provided selectively and variably by the state, with free power indirectly leading to free borewell water for the second community. These allowances were created, in the first case, through negotiations between governmental and nongovernmental departments that did not involve community residents except through their initial payment for the DEWATS. The leadership was weak and could not negotiate a beneficial social contract for water or sanitation. In the second case, local leaders represented the community in garnering consistent free power for groundwater pumping, to keep groundwater unpriced and much cheaper than tanker and ATM waters. The second community could wield power through its leaders to negotiate the provision of free power, and their common religious membership enhanced the solidarity hoped for in the social contract. Some of these features will appear in the third community case as well.

IN-DRAIN TREATMENT

Situated alongside the Yamuna River in the state of Uttar Pradesh, just across from the magnificent Taj Mahal, lies the third community. Its houses were situated adjacent to one another, clustered together but surrounded by open land and agricultural fields. The houses lined either side of the narrow lanes that navigated between them. The lanes, paved with large stones, were less than one meter wide. Open, narrow drains were running along both sides of the lanes and just outside the homes. In some places where the narrow drains carried a flow of wastewater, they were covered with removable concrete slabs. These narrow drains ran from all corners of the community into one large storm water drain that emptied into the Yamuna River. These small drains networking into the main storm drain formed the catchment of the sewershed for this community of nine clusters or *mohallas*. At the time of our survey work, there were approximately 450 households of 2,300 forming the community.

The Agra Nagar Nigam or the municipality of Agra estimated that in 2017 the black water flow in the main storm drain had reached 300 kiloliters per day (kld). At one time, a DEWATS facility built on the main stormwater drain was one of the most reported and talked about DEWATS projects in India. It was constructed by CURE, an NGO working in low-income communities, in cooperation with the London Metropolitan University and the Water Trust of the UK.[17] The project was part of the Crosscutting Agra Project (CAP) which was initiated in 2005 and supported by the Agra Nagar Nigam (ANN) and the US Agency for International Development (USAID).

The aim of the multi-stakeholder project was to improve the quality of life for low-income community members by providing better access to sanitation and sustainable livelihoods through tourism.[18] The objective was to improve the quality of the environment and use the treated water for irrigating the adjacent farmland. The unused treated effluent would be discharged into the Yamuna River. Constructed directly in the wastewater drain, the DEWATS was supposed to treat fifty kiloliters of wastewater per day. With no energy requirement, the system was supposed to bring down the BOD [biological oxygen demanding content] of the wastewater from 300 mg/liter to below 30 mg/liter using gravity flow, anaerobic underground tanks, and bioremediation with plants.

Lying in the middle of the storm water drain, the primary treatment system had three chambers (a screen chamber, a pre-process filter chamber, and a baffled septic tank). A secondary treatment system using an up-flow type of baffled reactor chamber filled with gravel completed the filtration of the water. A tertiary treatment system using a root zone treatment chamber of kardal canna and its roots further filtered the water. The treated water was supposed to be stored in an underground sump.

Our team of researchers made three field visits to this site over a period of one year to carry out interviews, engage with focus groups, and administer a survey created by my colleague Sukanya Das at TERI School

Figure 6. In-drain DEWATS. (Photo by Kelly D. Alley)

of Advanced Studies. This survey was created specifically for this community and asked questions on demographics, awareness of wastewater problems, the in-drain DEWATS, and opinions on its future sustainability. Our intention was to assess whether the community perceived that the DEWATS facility was meeting the intended objectives. We expected to document acceptance by the community. The survey created an analytical hierarchy process framework that summarizes the responses below (see diagram below).

In the community, the majority, or 82 percent, of households had an intermittent supply of piped, non-potable water from the local municipality, the Agra Nagar Nigam, which was provided to them for one hour two times a day. Residents also obtained potable water from an RO filter that community members installed in a shed several years earlier. The vendors sold the filtered potable water to other residents for Rs. 5 to 7 for a twenty-liter can. A smaller group of residents drew groundwater from a communal submersible pump and others procured water from hand pumps. Those receiving the non-potable water supply from the government said that they

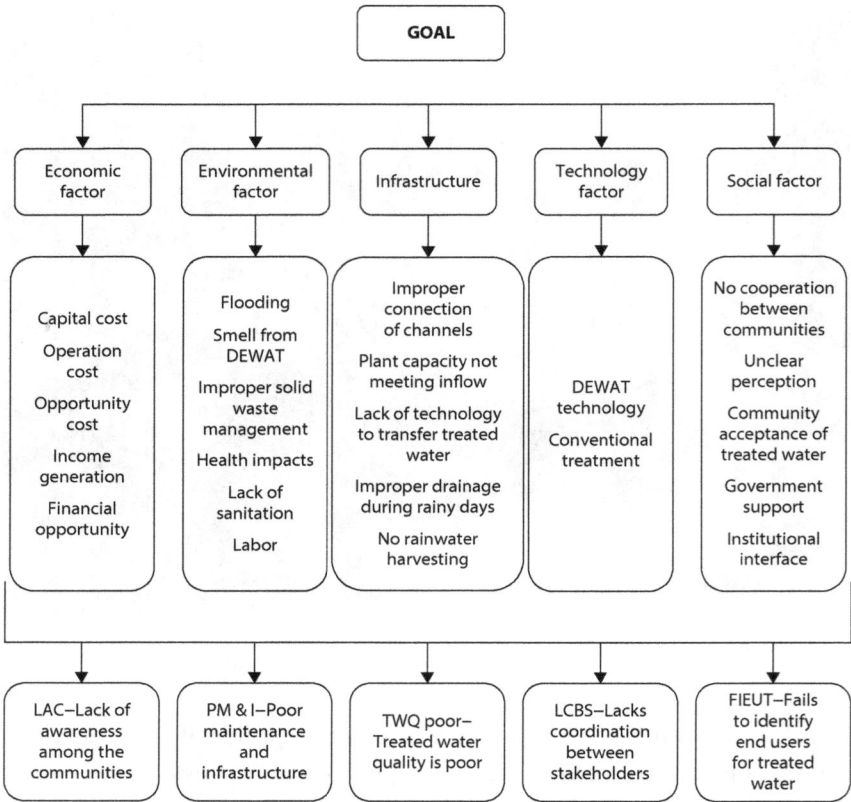

Figure 7. Factors and summary responses to survey using analytical hierarchy method. (Created by Sukanya Das, Shubhangi Chadha, and Shreya Annie Mathew)

did not pay any water bills. This was the government's social contract with them, free but limited, intermittent piped non-potable water.

One of the key sanitation objectives under the CAP project was to create a public toilet complex and support the installation of household toilets.[19] A public toilet complex was constructed one hundred meters downstream from the in-drain DEWATS, but it was destroyed over time. Some residents said that "antisocial people" had stolen the toilet seats, water tanks, and tiles. In response to our questions on the public toilet complex, residents were not aware of who was responsible for taking care of it or guarding it, and the premises degenerated into a single shell building with nothing inside.

CURE provided financial help to sixty-six households in the community to install toilets in their homes. We found that 59 percent of residents had toilet facilities in their homes and three-fourths of those with toilets had them connected to underground septic tanks. The other 25 percent with home toilets had pipes that carried the wastage to the narrow drains outside their homes. Even with toilets in many homes, half of the respondents claimed they had family members who engaged in open defecation.[20]

During the year of our visits, the DEWATS facility was maintained by an employee of CURE. He was responsible for preventing the drains from clogging by removing solid waste from the inlet point of the STP and from the overflow drains on either side of the reactor chambers. The community was aware of the STP, but only half of our respondents were able to explain how it functioned. About 10 percent of the respondents who understood the function of the facility claimed they never saw the cleaner water it produced and therefore found the DEWATS useless. Others explained that as the clearer water exited the STP it mixed with the black water flowing along the sides of the bioreactors and got absorbed by it. Most residents said that they had never used the treated water.

There were also problems related to the maintenance of the DEWATS. People living near the drain said that during the rainy season the drain overflowed and flooded the low-lying areas of the community. Some believed that the flooding was due to obstructions in storm water flow created by the in-drain DEWATS. The reactors situated in the middle of the drain were blocking the flow of rainwater. Solid waste thrown into the drain was also blocking the sewage and rainwater flows which caused a backflow of wastewater and solid wastes into the adjacent lanes. Residents said that the flood waters brought human excreta into homes in the low-lying areas.

Most of the residents said that if CURE could address these problems, they would be grateful. A quarter of respondents said that they had complained to CURE, and a fifth said that they had complained to the municipal corporation. A third of our residents wanted to remove the DEWATS from its present place and relocate it somewhere downstream, beyond the residential area. Twenty-one percent of residents believed that the DEWATS facility had improved the living environment of the area. They saw it as a place to organize functions. The concrete covering over the chambers in the drain could be used as a play area for children and a place for discussion and chatting. We found that CURE had invested in capacity building for some of the community members, but that the organization was not able to keep them associated with the drain work. Over time, community members grew disconnected from the NGO and the project. Some residents explained that when foreign delegations visited, CURE representatives would call a meeting to provide information about the DEWATS so that they could talk with the visitors.

Although ANN was a partner agency in this project, the representatives of CURE blamed the community for abdicating their responsibility. The representative of CURE claimed that at the time the DEWATS was constructed, there was a pact between this NGO and ANN which agreed that ANN would take over the operation and maintenance of the system after construction. Given this, the NGO blamed ANN for avoiding the agreement. ANN representatives claimed that they did not have the skilled human power to take care of the system or the funds to recruit someone. With only three sweepers assigned to the entire community, ANN could not keep the solid waste under control, and it mounted in the drain, partially blocking the wastewater flows to the reactors. In 2018, CURE decided to exit the situation and handed over the DEWATS facility to ANN.

Although the project made a marginal improvement in the quality of the wastewater flowing in the drain toward the Yamuna River, the management did not function as planned and the facility degraded over time. With the increasing availability of water, the volume of wastewater increased. Over time, the fifty kiloliters of treated wastewater was absorbed by the greater flows of black water running along the sides of the bioreactor. The downstream mixing of treated with untreated wastewater discouraged people from identifying the treated water as usable. With the only benefit being a place for congregation over the drain, the community was not interested in looking after the treatment process. They invested nothing in it and the plant generated no usable water. Local government support was absent.

MINIMUM SOCIAL CONTRACTS AND THE CUNNING STATE

These three cases exhibit instances where residential communities have been recipients of social contracts with state and nonstate agencies but have generated no substantial collective action around sanitation. The social contracts have covered water or energy supply, and the sanitation social contract was provided through the creation of infrastructure by arrangements between governments and NGOs. Although the government intentions, stated in policy documents, were that the STP operation and maintenance would be carried by state or NGO agencies, there was less than a bare minimum contract for the STPs. The STPs looked like they worked but they did not, and they produced few benefits and no reuse water for the communities.

Chidambaram has also argued that in low-income communities, there is more collective action around water procurement than sanitation. Communities coordinate better around club goods than they do around public goods, but she pointed out that not all public goods can be converted to club goods.[21] Deeper structural issues such as centralized infrastructures are obstacles to collective action around sanitation. Toilet and STP construction involve greater infrastructural adjustments, land, and collective planning than water provisioning does, which involves government pipes, ATMs, tankers and groundwater and hand pumps. While the cases have shown that land is constrained, the actions of STP users proved that decentralized infrastructures can be invented. STPs were created adjacent to blocks of apartments in the first case, within the settlement in the second case, and within a storm water drain in the third case. The placements were all innovative. However, the cases show that operation and maintenance of the facilities were bigger challenges because communities were not directly involved in financing the STP or perceiving benefits from its operation. In the first and third cases, the communities received intermittent and limited supply of non-potable piped water from state governments, which barely ensured this public good. To meet their full needs, they had to spend time and energy securing additional waters of different gradients (potable or non-potable) from a variety of providers, arrangements that all fell outside the social contract of the state. At the same time, all three communities suffered from tangled relationships with state and nongovernmental agencies in the delivery of sanitation services. The communities were not able to play a role in the initial decision-making on technology and design, or in upgrading and sustaining the projects. In these three cases, STP projects were languishing or defunct and residents were not able to engage with them in ways that benefitted them. There was very little engagement and no experimentation, and therefore little interest in generating reuse water.

The draw-down created by the tangled governmental-nongovernmental arrangements involved the cunning postures of state and nonstate actors. This frustrates the more optimistic argument that public goods should be coproduced by communities, and governmental and nongovernmental agencies including those in the private sector.[22] While coproduction can bring together resources from multiple parties to assist in sanitation for the urban poor, there are real problems if urban residents are not in charge of decisions and user options and benefits. In these cases, the agencies involved in constructing and handing over projects displayed an on-again, off-again approach to their work. The STPs worked, but then they didn't. The responsibility to repair and maintain them was passed from one agency to another without any resolution. In these passages, the state was not the only cunning actor. Nongovernmental organizations also appeared cunning in their collaborations with governmental agencies during the initial construction of the STP and in the pretenses to maintain them. While the intentions appeared good, the follow through work in bringing communities on board to sustain the projects was weak. Over time accountability waned and projects languished with no real owners. In the cunning relations among state and NGO agencies, the STP became a pretense, a stage on which outside agencies pretended to perform important functions. Consequently, the communities found the projects rather useless and ignored their functions and potential to produce reusable water.

In the next two chapters, I show how communities and colleges are finding ways to manage the double burdens of water and wastewater services, but they face significant costs. In the following cases, community members and state and city leaders will advance beyond rudimentary social contracts to establish new kinds of arrangements with private companies. These arrangements and contracts will move communities into experiments with reuse options.

4 Horticultural, Partial, and Off-Grid Reuse

As small-scale wastewater treatments were being invented and optimized over two decades, private companies, institutions with large land complexes, resident welfare associations (RWAs), and nongovernmental and civil society organizations entered into agreements and arrangements. In these arrangements, they were starting to generate reuse experiments and options. Beginning with reuses for horticulture and gardening, some partnerships were developing other reuse ideas. Generally, in middle- and upper-income apartments, bathing, drinking, cooking, and dishwashing take up 40 percent of the water budget, while toilet flushing, household washing, outside and building cleaning, AC cooling, and gardening take up a bit more than 50 percent. Incrementally, these nonessential needs, which account for 50 percent of the water budget in urban areas, were being met by treated wastewater in some housing communities. As users were finding, treated wastewater could help save groundwater and freshwater supplies for essential needs. Recycled water could also provide backup water for disasters and emergencies.

This chapter and the next provide examples of partnerships in treatment and reuse in which municipal leaders have promoted new schemes and communities have drawn new relationships with private companies and NGOs in STP installation and maintenance. I divide the projects in chapters 4 and 5 by their levels of reuse, into the following categories: (1) full-fledged horticulture reuse, (2) partial reuse, (3) off-grid reuse, (4) closed loop reuse, and (5) emerging reuse. Within these types, I introduce project decision-makers and outline some of their decisions. I explain the partnerships they create or work with, and some of the issues they face implementing and maintaining technologies to meet their needs. I introduce their perceptions on the behavior of microbes in treatment processes and water storage. Unlike the cases in chapter 3 where involvement of governmental and nongovernmental agencies was initially strong but community engagement was almost nonexistent, these cases show that municipal leaders, communities, NGOs, and companies are interacting through a variety of agreements.

I have selected the cases for chapters 4 to 6 from two hundred projects I visited alone and with student teams over a span of four years. The recycling projects were housed on university and hospital campuses, in housing complexes, and at airports, malls, and city parks. I work with a flexible understanding of how residents use these technologies over time, drawing from Starkl et al.'s approach called "flexiBAT."[1] FlexiBAT is an approach to the assessment of best available sewage treatment technologies that employs flexible definitions of functionality, success, and failure depending on the context and specific technical and institutional conditions. Starkl et al. compared the costs and benefits of several small-scale technological systems; some of these technology types were described or mentioned in chapters 2 and 3.[2] They include baffled reactors and soil biotechnology reactors, and the phytorid bioreactor I will introduce in this chapter. Aeration devices such as the vortex and membranes for filtration, including soil and carbon filters, will appear in these project descriptions. The systems incur different capital and operational costs (called capex and opex in business parlance) and require varying amounts of technological and regulatory knowledge to operate and supervise. In their comparison, Starkl and his team found that some technologies were more costly to operate than others. Some were better at removing pathogens than others. While I do not assess the effectiveness of these technologies, I will elicit stories about the decisions and choices companies and communities make to set up, upgrade, optimize, and finance their projects. I use the flexiBAT approach to think flexibly about how projects are produced by their own unique circumstances, without dismissing shaky experiments with the discourse of failure. I will also argue that something that looks like failure may not be complete failure, and that some measures work better than others or better than none at all.

The government agencies appearing in these cases are the water supply and sanitation boards, namely: the Delhi Jal Board (DJB), the Bangalore Water Supply and Sewerage Board (BWSSB), the Brihannmumbai Water Supply Board (BWSB), and the Chennai Metro Water and Sewage Board (CMWSB). The nongovernmental agencies that are mentioned are those introduced earlier, the CDD, NEERI, the CSR in Auroville, and Ecoparadigm. For the range of existing projects that I do not mention, there are other private technology and service maintenance companies. Project operators must answer to the main regulatory agencies introduced in chapter 1: the Supreme Court, the High Courts, the National Green Tribunal, the central and state pollution control boards, the Central Ground Water Authority, and the regulatory arms of the city water and sewer boards. The agents connected with the pollution control boards do site inspections of STPs, collect reports on STP functioning and water quality parameters, and publish lists of labs qualified to analyze water samples taken from the STP effluent (after the last phase of the treatment process). They enforce bans on groundwater extraction by penalizing or closing down facilities.[3] The state pollution control boards are the primary monitors in each state but other consulting agencies, such as the CDD, Ecoparadigm, and NEERI, provide assistance in monitoring, design and optimization.

FULL-FLEDGED HORTICULTURAL REUSE

I look first at the projects dedicated to horticulture reuse because these are the most abundant across the field of decentralized projects. The first set of projects was set up by a group of political and administrative leaders within the Delhi legislature and the New Delhi Municipal Council to provide recycled water to the beautiful gardens of New Delhi. The chief minister of Delhi (which is a union territory rather than a state) and the chairman of the NDMC (one of the municipalities in this union territory) has pushed for "lifeline" water and access to water ATMs and sanitation facilities in public areas. To ensure more control over this plan, Chief Minister Kejriwal took over the water portfolio in the Delhi government in 2017 after initiating reductions in tariffs for household water supplies.[4] The tariff structure in 2015 allowed households to consume up to twenty kiloliters of water for nominal charges, which he then referred to as "lifeline water." But after a short time, the DJB protested that the city could not afford such a gracious provision, and the DJB reset the volumetric rate at a slightly higher level to recover some revenue from households using less than twenty kiloliters a month.

The New Delhi Municipal Council manages the region that houses the central government offices, the embassies and consulates, and other entities involved with national security. The NDMC gets 124 million liters of potable water each day from the Delhi Jal Board and pays Rs. 15 per kiloliter. Borewells and rainwater harvesting meet the rest of the NDMC's need. Borewell water is generally priced at about Rs. 10 a kiloliter to pay for the electricity costs of pumping. The NDMC distributes the 125 mld of water from the DJB through its grid of freshwater pipelines. It also sends tankers full of water to communities that live off this grid. The NDMC prices this water according to a sliding scale that gives the first twenty kiloliters of water to households at the rate of Rs. 3.45 per kiloliter (the nominal rate revised by the DJB) and then prices the water volumetrically above that.[5] This means that households with lower usage are subsidized by the higher rates charged to big volumetric users such as five-star hotels and industries.

The NDMC region, once called Lutyen's Delhi, is well known for its historic buildings, beautiful gardens, and monumental landscapes. About 80 mld or 10 percent of the NDMC's combined water budget used to be allocated to the upkeep of the city's eight thousand parks and gardens, and for fountains and lakes. These facilities have no user fees associated with them. Up to 2017, this 80 ml of water was cobbled together each day with NDMC water from the DJB, referred to as "pure" water, with groundwater and with treated wastewater from a large centralized STP at Okhla.[6] The centralized STPs in Delhi have been treating around 455 million gallons every day (mgd) but using only a fraction of that for reuse. At the time of my research, they claimed to be using 142 mgd (or 645 mld) of the treated water for horticulture and irrigation in the metro region, but I was not able to verify this with my research colleagues.[7] We were able to corroborate that the treated wastewater from the STP in Okhla was being used for the NDMC parks through a pipeline we could never fully trace. But it is possible to find water occasionally bubbling up from individual manholes on the streets near the monuments and VIP homes in the area and I could conclude from the smell and the fact that it looked clearer than raw wastewater that it could have been treated water from Okhla.

With groundwater levels depleting to three hundred feet below surface level in some sections of Delhi, the NGT issued an order in 2017 that all urban municipalities must curtail their use of groundwater for horticulture and use treated wastewater instead. Before this in 2016, the NDMC had taken the decision to promote decentralized STPs to deal with the wastewater load in the city and to promote recycling of treated water for horticulture and irrigation. The usable water produced by these decentralized plants combined with the minor supply from the Okhla STP could be used for the city gardens. The NDMC installed eight decentralized STPs within the year of 2016–2017 and proposed to develop ten more over time. Reducing the dependency on groundwater, these projects produced treated wastewater from the raw wastewater drawn from underground sewers leading to the centralized STPs. To connect these decentralized STPs to the centralized sewer grid, the contractors only needed to construct a short pipeline which in all cases was less than 750 meters. This initiative grafted a decentralized solution onto a centralized grid.

At the outset, there were two private companies working with the NDMC to construct and maintain these new STPs. One company used the Soil Biotechnology bioreactor invented at IIT-Bombay (introduced in chapter 2) to build five STPs in different gardens. They used local labor and consultants from the private company the inventor had created. These STPs used five hundred square meters of land for a 500 kld plant and slightly less land for the smaller 300 kld plants; this is a fraction of the land required by conventional centralized wastewater treatment plants. The NDMC also hired a second company to build three STPs using the MBR (membrane bioreactor) technology. The MBR technology also requires very little land but involves aeration to stimulate bacterial multiplication and thus uses more power. The unit contains an initial chamber where grit

Figure 8. Author taking a picture of treated wastewater being filled into a tanker for garden use. (Photo by Nutan Maurya)

and solids are removed and a tall metal tank where the water is then aerated. In this open-top tank, the wastewater is provided with a continuous supply of oxygen to spur the multiplication of aerobic bacteria, which then digest the biomass in the water. The clearer water runs off from that tank and through a membrane filter housed in an adjacent room. This membrane filter removes remaining bacteria and particles and produces water usable for horticulture. Sludge is generated from this process and must be removed from the tanks and distributed for horticulture use. The treatment capacities of these SBT and MBR projects are in the capacity range of 100 to 500 kld.

In the public–private arrangement between the NDMC and the companies, the companies were to cover the costs of building and maintaining the STP for twelve years. These costs included laying pipelines from the nearest wastewater sewer to the plant. The installation costs included financing, construction of the tank and pipelines, materials, operation and

Figure 9. Nutan Maurya at the rose garden with an SBT plant in the foreground on Shantipath in New Delhi. (Photo by Kelly D. Alley)

maintenance costs, staff salaries, water storage, and distribution mechanisms. The companies did not incur any land costs, as the NDMC provided free land within each park. In the contracts, the NDMC allowed access to the city wastewater drain where the raw supply was drawn and agreed to buy all the water produced by the STPs. The NDMC distributed the treated water through the garden water pipelines and used its own water tankers to collect the remaining water for distribution to other gardens and roadside medians in the union territory. The NDMC agreed to pay between Rs. 30 and 37 per kiloliter to the company for water on the days that they collected and then added a lower-rated payment for those days that they could not collect the water due to weather or other circumstances. The incentive for the private company was that it would get regular payments from the NDMC, pay off its investment in five or six years and glean a profit from the regular payments made by the NDMC after that. The contract relieved the government of having to come up with the

Figure 10. MBR plant in the NDMC area. (Photo by Kelly D. Alley)

initial capital for these projects, something that is generally difficult for municipalities to manage. But the NDMC had to provide consistent payments for the treated water. The government entity became the consumer of treated water rather than the producer, and the producer company was able to propose the technology and maintenance mechanisms in consultation with the NDMC.

Lloyd Owen reported in 2016 that water reuse is a small element of public–private partnership contracts, but the percentage has been increasing since 2005.[8] He found that the contracts are usually awarded in middle- to high-income countries in areas with high water stress. While PPP arrangements in water provision are highly contentious in India due to fears and the actualities of higher water prices, PPPs in sanitation receive very little attention or interest. This particular arrangement is one of many that are being explored by the Government of India. Other types include PPPs in which an existing STP, constructed earlier by a government entity, is loaned out to a private company for upgrading, operation, and maintenance. In the decentralized field, however, there is a different progression of activity in cases involving housing societies. First the builder of the housing complex builds the STP as part of initial construction and then hands over the dysfunctional unit to a resident housing society or RWA for upgrading, operation, and maintenance. Many RWAs hire a consultant and contractor to upgrade the system (or at least make it functional) and then hire a manager for operation, maintenance, and meeting regulatory requirements. These kinds of arrangements will be seen in the following cases in this chapter and the next.

PARTIAL HORTICULTURAL REUSE

In Mumbai, I found a women's college that had established a scheme to use treated wastewater for gardening before the NDMC projects were built. The college had adopted a phytorid technology created by NEERI, the National Environmental Engineering Research Institute.[9] The phytorid technology combines a baffled anaerobic reactor and a planted filter. Once the wastewater enters the baffled reactor, anaerobic bacterial digestion occurs as the water moves through several chambers. As the solids sink to the bottom of these chambers, the clearer water is guided out of the reactor and then pushed horizontally through the planted filter. This technology builds upon the planted filter designs that the CDD and the director of the Centre for Scientific Research in Auroville described in chapter 2. As one of the monitors of this phytorid plant explained to me, "You don't add any bacteria to it. It feeds off itself without oxygen [within the baffled reactor]. There are vents for gases to exit. Solids get trapped in the first chamber. The gravel section is three meters deep and there is a horizontal flow through graveled baffles. The staff can dislodge sludge when it gets clogged in the gravel by flushing it out with jets from access points in the filter."

The original anaerobic bacteria in the wastewater starts multiplying in the baffled reactor in the absence of oxygen and then eats the biomass. It continues to do so as it moves through the planted filter. Unfortunately, on the day of my visit to this college, things appeared very messy. I kept asking questions: Why isn't the water flowing through the planted filter?

(I saw that sludgy water was collecting on the surface of the gravel.) Why is this manhole for the baffled reactor lying open? The guide from NEERI explained through my ignorance that various repairs were underway. The maintenance company, a privately run entity, was handling this 100 kld plant and was planning to build a 100 kld addition on the request of the college. On that day, the wastewater was not being treated but was being bypassed directly into the drain (nala) running beside the STP. Nearby and within sight of this STP, a large office building housing the Merrill Lynch company, was dumping its wastewater directly into the same drain. So, I asked, "Do these large office buildings have the same requirement to install STPs and reuse their treated water?" No, he replied, "They understand that they should do it, but it is not required. It is not said."

I continued to ask about the college's financial situation with its water supplies. By using the treated water for gardening, the college was able to save the money they were paying out for water tankers. Before the start of the STP project, the college was procuring sixty-five tankers of water every day for gardening. When switching to use of the treated water, they were able to reduce their tankers to forty every day. They were receiving piped water from the municipality for their drinking water needs. I asked about where the tanker water was coming from and the representative replied, "We don't even know the quality. It costs Rs. 700 to 1000 per tanker and we don't know where it comes from. The Government doesn't give tanker water for gardening." He went on to tell me that the cost to upgrade the STP with an additional 100 kld in capacity would be Rs. 20–22 lakhs (2,000,000–2,200,000). To create a 200 kld plant, the total installation costs could reach Rs. 40 to 45 lakhs (4,000,000–4,500,000).

In each place I visited, I tried to collect estimates on costs of STP installation and maintenance and on tanker and ATM water prices. In Delhi, I was told that the cost of a 300 kld SBT was around Rs. 65 lakhs (6,500,000). In part, the costs depended upon where the building materials could be obtained. If local materials such as gravel and brick pieces were available, the costs could be lower. If the materials were imported from another region of the country, then the costs ran higher. As I was documenting information on installation and operating costs, the data pieces started to sync into rough averages. Such synchronizing of piecemeal data became a real asset given that most private companies were not

Figure 11. Phytorid plant at a hospital in the Mumbai metro area. (Photo by Kelly D. Alley)

willing to provide copies of their invoices and receipts to show the exact accounting of their expenses.

I employ the label of "partial" reuse for this case and others because I noticed that some projects treat wastewater and generate reuse from *time to time*. Rather than consider them part-time failures, I suggest that there are cycles of functionality. When an STP has just been completed, is running, or repairs have just been made, then things tend to look good. But inevitably, over time, a pipe gets clogged or a tank gets filled with sludge or silt, and the process must be halted. At that time, a "honey-sucker" or sludge removing company—usually a private company—is called in to suck out the sewage. This honey-sucker ends up disposing of the sludge in another way, by dumping it on vacant land or into a drain somewhere else.[10]

At another project that I put in this "partial use" category, a Mumbai-based builder was installing STPs in his housing complexes at the time of construction. After construction, his company was maintaining the STPs for residents. The builder explained to me that he maintained the facilities because the residents would not. With numerous projects in various stages of development, he had to show operational STP status for existing complexes before applying for permits for new projects.[11] "Do other developers do the same," I inquired? He responded, "They operate for 'namesake'" which I understood to mean that they pretend to operate or say they operate but do not. He continued, "We have to do the regulations now to continue to stay in business. It is mandatory by Centre regulations." Regarding others who may not have been operating their STPs, he remarked, "They just skip those things and do that" [in non-compliance]. As for his own experience, he claimed that the authorities being very strict, he had to submit reports to the state pollution control board every six months and maintain them to get approvals for other projects. He claimed that the PCB came and assessed the projects periodically.

Since the builder's company was operating in place of a Resident Welfare Association (RWA), I asked him whether they saw any savings in the water bills by using the STP water for gardening or other uses. He replied, "We don't save on water but there is a safety part. We know we have a backup and have water in the summer." In the housing complex where I took his interview, the company was using borewells to extract groundwater. But the high TDS (total dissolved solids) in the borewell water prevented residents from using that water for household purposes. His company used RO (reverse osmosis filtration) to filter the borewell water in other buildings, but they could not in this complex due to the high TDS levels. During the peak summer months, they were able to use their recycled wastewater for non-potable uses and save their potable water supply for essential uses such as drinking and cooking.

In terms of machines and operations, they have had to adjust. At the time of our meeting, he told me that they had recently switched out their MBBR bioreactor [moving bed bioreactor], which aerated wastewater in a tank under the parking deck and replaced it with the soil biotechnology bioreactor. They found that the operational and repair costs of using air blowers to oxygenate the wastewater in the MBBR technology were too high. They switched to the soil biotechnology method because it was cheaper to operate. There were fewer mechanical parts required for the SBT method and the mechanical knowledge required to run the unit was minimal. He explained, "Power for pumps and blowers was too much in the other technology. You are using blowers and sometimes the blowers get choked up. In this new bioreactor you can see the pipes, so you don't have to get into the sewers or sewage that much." In the MBBR technology, when silt accumulated in the tanks, they had to call the municipal corporation or someone with a tanker to pump it out. He added, "In our new [SBT] technology, there is no sludge and it [the bacteria-laden sludge] grows slowly so we don't have to extract it at all from the tank. But if there is silt and suspended solids, then it won't dissolve, and you have to clean it out."

During my journeys, I found other partial reuse cases in which the resident communities were more involved. For example, an Army residential colony outside Bangalore was planning upgrades when my research team and I visited the location a year later. While interviewing a retired army officer in this residential colony, we found out about the community's decisions and choices. He explained that his house lay at the lowest point in the complex and that during heavy rains, the water flowed down to his part of the neighborhood, where they had created six rainwater pits. They created a catchment system so that all the water flowed into these pits through the storm water drains. Each home also directed rainwater into individual pits in their yards. They lined the pits with charcoal to manage the percolation into the ground. If the household pits overloaded, then the rainwater drained into the neighborhood pits. All these systems helped to recharge the groundwater. He gave us some background to how they developed these mechanisms off the grid of the centralized water system:

> When we moved in five years ago, we did not have connectivity with BWSSB [the Bangalore Water Supply and Sewage Board] or Cauvery water. We don't have water lines, so we survive on groundwater, and we extract it through borewells. But most of them have dried up. We have a 100-year-old borewell and we were even sustaining coconut groves at one time. There used to be lakes, but they have disappeared with construction. Today we are still using this old borewell. When we moved in here, there hadn't been rains in the past two years. The builder of our homes and neighborhoods (called pockets) made the rainwater harvesting but we did more and did channelizing and check dams. We did our own homework and we diverted [rainwater] to these pits. We don't have waterlogging now as it used to happen because we applied our mind and diverted all the water to the pits. So, we were successful in recharging the groundwater here. From here, we pump the water underground to a reservoir all night and in the morning, we release the water to all the pockets. We are building a water treatment plant at

the reservoir. In each home, we have our own RO system. We need to clean the weeds and algae in the pits once in six months with alum and chlorinate it. Then it becomes so blue in color. It costs a lot, but we do it. So that is drinking water. The STP is another topic.

The community reservoir combines groundwater and rainwater, but this is not used for potable purposes. Sometimes, he said, they get tanker water during dry spells in the summer. They were pumping the reservoir water on alternate days for half an hour to each house. Each house was getting three thousand to four thousand liters during that time every other day. They were storing it in their overhead tanks and sump tanks.

Turning to the STP, he explained that their project was shared with three pockets. In one pocket, there were 112 housing units, in another there were 32, and in the third there were 70. All the sewage lines from these pockets were running to this single STP. The builder of the complex provided the original STP, but the residents decided to upgrade the system. He explained,

It is being done. We have done punch listing and snag listing and it is under modification by Dr. XXX. He is a specialist in STPs, a known figure. He is modifying the STP that the builder built because the capacities do not match from what the builder did. The aeration tank is supposed to be inside, but it is now outside. It is supposed to have more area, to have more exposure to the air. So, we are building another tank outside to store the sewage that comes in and it has screens and bars. The aeration will occur in that new tank. The current tank we are using was shoddily built but we are fixing it. A project director is fixing it. After the aeration, it [the wastewater] goes through a sand filter and then they put some chemical in it so it doesn't stink. It turns out looking pretty good, but we use it for gardening in the common areas for the entire complex. We are generating so much treated water that we can sustain all the pockets [beyond the three mentioned here]. The other STP in the other side of the community doesn't work.

They paid a monthly maintenance fee of Rs. 3,200 a month to cover water charges, waste collection, security, and sewer maintenance. He mentioned to us that the homes in his community did not have dual plumbing to enable toilet flushing or gardening with treated wastewater. But he was still conducting recycling within his own home, reusing the household gray water and the RO wastewater for his garden. For drinking, he said, he was not convinced [to use the treated wastewater] unless he knew about the technology. Treated wastewater was OK for landscaping and washing cars, but he didn't have faith in the system managers to follow best practices at the time of treatment. He continued,

> I would love to be convinced for drinking, but so far I am not. . . . It shows my lack of awareness . . . as an army man I used to drink, let's be very honest. But I trusted the best practices being followed. We believe in zero tolerance in our houses. In anything. That is the best practices. It is followed to the T. . . . I had faith. But here in this system, I don't have faith. I am sorry to say in a first world country I may. We still have to come of age. That is a challenge in the society in which we are living. . . . I would love to say yes, but I can't say yes. It all depends on the commune in which you are living. There has to be monitoring. We haven't reached that stage as yet.

In his community, the housing society or RWA, not the municipal board (the BWSSB) or the PCB, regulated water uses and STP operation. The RWA members were choosing the technology upgrades and financing them in their off-grid location. The retired army officer mentioned that they had no connection to the Cauvery River, which is the primary freshwater source in the centralized water system of the Bangalore metropolitan region. If a community has a piped connection to the Cauvery River water, then a sense of security and well-being can be achieved. Without it, the community must take charge of all their water needs and search for sources, as this army colony has done. This sentiment about the centralized Cauvery connection appears in the next example of partial reuse and furthers the understanding of the disconnections as well as interfaces of decentralized and centralized systems.

PARTIAL REUSE AND THE CAUVERY CONNECTION

In some settings, partial reuse is reconfigured when neighborhoods become connected to a centralized city water grid and start receiving greater amounts of municipal piped water. On the other side of Bangalore, the manager of a large complex of six pockets explained that one of their neighborhoods or pockets was using water piped from the Cauvery River while the others had not yet received a connection. This Cauvery connection led them to make different decisions for water and sewage management. The difference occurred because the six-pocket colony was straddling the boundary of the city limits of Bangalore, and one pocket was within the BBMP water district and the other five lay outside the district on rural, agricultural land. The pockets located outside the city district had no connections to the piped water grid. I will explore resident views in the pocket connected to the Cauvery source and then turn to the other pockets where residents live in off-grid conditions.

As one resident explained, they got their "Cauvery system" after five years of living in the housing community. Before this connection, they had two active borewells in their pocket, but their water was hard. So, they had to run their borewell water through an RO sodium-based plant for softening before they could pump it to the individual houses. The fifty-seven houses in their pocket shared one tank to avoid needing individual storage tanks at each house and they received pressurized water from this centralized tank. There were no flush tanks in their bathrooms. A resident explained,

> We are using Cauvery water right now and that water comes treated. So we don't need to treat now [their borewell water]. We have three tanks, each of 2 lakh (200,000) capacity. We haven't used our borewells, as we have Cauvery water. But Cauvery water is more expensive than borewell water. We are a community based on sustainable living and we practice certain things as a group, such as STP operation and biogas conversion, composting, and waste reduction. We have a board [an RWA] that takes decisions and runs these things. It is a collective responsibility. We have rooftop rainwater harvesting and we collect it on the roof, and it is fairly good water. Our tank fills up in rains before summer, so it is our backup in case the Cauvery supply doesn't come. We get supply all the time though, so we are fortunate. We have RO plants in our houses. We used our borewells only in the first five years, but in the last two to three years we haven't used them, as we have Cauvery. We use the STP water for gardening. Now we are under BBMP, before it was rural. That is why we get Cauvery water. The other pockets are still rural. There is a rule that you have to be operational for a few years. Land next door to us is still rural, agricultural land, unregistered land. We needed our own STP, more from consciousness than anything else. Plus, the BBMP has the rule that we should have our own STP. These are new rules that are not implemented completely. It is better to be self-sufficient. We are happy that we are not putting waste in the river.[12]

This respondent was aware of the water system and the trade-offs in the community. She was also aware of the regulations and requirements.

Figure 12. A planted filter in an upper-income housing society. (Photo by Kelly D. Alley)

Likewise, we found that residents with active RWAs around the country had a high level of awareness of their water and sanitation systems and engage in several ways with local and state governments.[13] In this response, the supply of Cauvery River water influenced their decisions on uses of other sources such as groundwater. In this pocket, the community had a large, planted filter situated at the edge of a nearby river that I noticed was overwhelmed by wastewater from upstream drains. They reused the treated water for gardening in their pocket.

OFF-GRID REUSE WITHOUT THE CAUVERY CONNECTION

The uses of treated STP water appeared more varied as we explored responses of residents in the adjoining pockets. We talked at length with the estate manager hired by all the pockets to look over water and sewage. He said that the other five pockets (which consist of at least one apartment building or an assortment of villas) have their own STPs and have their own flushing lines. The builder installed separate lines, one for STP water and the other for potable water, during the construction phase. All the pockets were able to add rainwater into the STP water tank during the rainy season. Rainwater was also added to the treated wastewater storage. By doing that, he said, you get a reduction in smell. "You want to store [the treated water] but the bacteria increase after a few days. So, I add chlorine. But I can use rainwater in the rainy season [for dilution] and reduce the chlorine costs." In another pocket, they use vortex aerators but do not have sand and carbon filters. They chlorine dose at the end of treatment in all the pockets.

Since the adjoining pockets of this community lie outside the city water grid, I consider them in the category of off-grid projects. Although all six pockets were within the same housing community, their conditions were very different due to the presence or absence of the Cauvery connection. The other five pockets were composed of several types of housing (apartments, duplexes, and independent villas), but they were all using the STP water for toilet flushing and horticulture. In 2007, the Bangalore-based company, Ecoparadigm designed and installed STPs in all these subunits. The consultant from Ecoparadigm explained about this community's units,

> It is basically a glorified septic tank. It gives time for the bacteria to work. An optimized septic tank. It works when it has to work, and we didn't want to make it more complicated by adding aeration. [I add that this is the only baffled reactor that I have seen that produces flushable water without other tertiary treatments]. Yes there is also a tertiary treatment since it is going to the flush. So we have aeration to remove the smell. You would have seen the two vortexes in the main office, but in other places we use a more traditional aerator and diffuser and a carbon filter, as you can't have too much suspended solids. Water is already softened in primary treatment. There is nothing centralized.

"So, is it a complete closed loop?" I asked. Nodding, he explained that there was a closed loop from the treatment tank to the flush and to irrigation. But it was not purely septic to closet. There was a certain amount of freshwater coming into the system. Water was added to every loop. The septic tank water goes back to the toilet. Dual plumbing was part of the original builder's plan. He was proud of the fact that they installed dual plumbing several years before it became mandatory for all housing complexes in 2015.[14]

I was hearing about the problem of smell from other residents using treated water for toilet flushing in this complex, so I inquired further. "Can you tell us more about what causes odor?" Regarding a current issue, he answered that a vent was improperly designed and was not taken vertically up. He added, "It should not be a problem but when it is near the parking lot, we have to move the vent around. The pipes can trap water and the fumes don't go out. So the gas gets [its] first release in the WC. So we redid some vents." He explained the maintenance and monitoring further,

> Here we have an issue with the filter media [the rocks or other kinds of materials added in the anaerobic tank]. Granuals have an issue here. CDD does concrete slabs and cinder, but other consultants prefer a PCB pipe. There was some clogging, and the bacteria was not working. We redid the STP and think it is not much problem now. It's not going to be crystal clear water coming out. To get that quality of water the investment goes up. We are looking at BOD of 10 in the outlet quality. Our monitor [for the RWA] checks for E. coli and checks the general functioning and when there is a smell. He checks the borewells too. Each subunit has one monitor on staff.

He explained that there were two settlers where solids are removed before the water flows to the bioreactor containing the media. The media, which were plastic floatable pieces, introduced more surface area for the bacteria to attach to. The greater surface area allows the bacteria to multiply and aids their digestion of the biomass in the wastewater. The media in these STPs were made of cinder or plastic pieces. The engineer mentioned that CDD, the design consultant, was visiting periodically to check the levels.

Another resident commented on the problem of smell within the context of her acceptance of the reuse system:

> We have no issue. We haven't had any health issue. But there is an odor issue at times, and we inform and then it is taken care. It smells like sulfur, and they add some chemical. It is rectified and solved. I wouldn't say the water is crystal clear. It is slightly brackish. Probably it is the quality of the treated water. We have to clean our toilet every day. It leaves a kind of layer, and it has to be taken care [of]. We have gotten used to it. For kitchen, washing,

and drinking we would not want to accept it. . . . It may be in your chemical analysis. It may be drinking water but I know what the source is, so I don't want to touch it. An expert here told me its drinking quality, but I don't want to use it. Would you four want to use it? [She points to us in the research team]. This is private and we don't know where the parameters fail on a daily basis. So we are not ready for that. . . . This is the first time I am using this water for flushing. It was not difficult to switch to this water. No, it was not difficult. I saw my sister using it in her villa in Chennai. I know there is an odor problem sometimes. In the other buildings, they have this problem more often, but if we get it, we immediately inform the maintenance.

Ecoparadigm maintained these STPs for the first few years as part of the original contract with the builders of the housing units in these pockets. Then in 2011, the community comprising all the pockets hired a manager and he started maintaining the STPs for all the pockets, which were in the capacity range of 40 to 60 kld for each. The average operation and maintenance (O&M) costs per year were Rs. 1–1.5 lakh (100,000–150,000) per STP for the planted gravel filter technology. These costs were borne by each pocket's RWA maintenance dues and the funds went toward sludge removal or desludging, vehicle charges, pipe cleaning, gravel costs, cement, staff, and motors for pumps. As the manager explained, "Maintenance becomes more after the first five years due to the need to replace materials. Madam; now the cost is high compared to five years back."

In this community, five of the six pockets were functioning with their own resident housing societies or RWAs. The most recently built pocket was still in the process of creating its RWA. Once created, the RWA would take over the maintenance of the STP from the builder or Ecoparadigm for that pocket. In one pocket, they had a nine-member resident committee handling different jobs involved with procuring water, landscaping, maintaining the STP and reusing the STP water for gardening and toilet flushing. They assisted the estate manager working for all the pockets. One resident explained how they worked with their STP:

RESIDENT: In between we had some problem with the flush toilets. Initially the sand filter had some problems, and now we are setting it right. Beginning of this year, we had some problem with smell, but it is OK now. For gardening we use the water, and it doesn't smell much.

KA: OK so maybe the sand [in the filter] was deteriorating?

RESIDENT: We have agencies that come and clean and we pay for it. CDD comes and does work for our STP.

KA: So you don't have to pay a contractor to maintain it?

RESIDENT: So far, we have not done it [pay a contractor]. We get someone [from CDD] to come when we have a problem. We have one estate manager here [the one for all the pockets] and we pay him out of our RWA maintenance. Another person looks after STP water. In the morning, he goes and sees the water supply and checks the STP water for gardening and flushing. He will check the quantity of water.

KA: Do you think it is easier having this system under the RWA or was it easier to have the system run by the original builder or contractor?

[GRA #1 Interprets my question again to the resident]

RESIDENT: The builder [through Ecoparadigm] maintained it for only two years. We have done it for the last two years. We have not faced much challenge.

KA: So it is not too much to handle?

RESIDENT: The managers [estate manager and resident committee members] are aware of the entire system, and they take care of it. They do level checks and arrange the water samples for reporting and testing and they submit the reports. CDD also takes the water samples and submits the report.

The resident then turned to discuss some issues with the STP manager. They were talking about the aeration tank and placing an order for the aeration pump. There was some R&D work going on and an issue with a bypass line. Turning back to us, the resident explained that the water-quality reports that they get from the CDD and PCB are circulated among the nine-member resident group that handles different tasks for the RWA. He explained that the PCB comes and inspects or the CDD takes samples, gets the analysis done by government-authorized labs and provides the reports. Then they file the reports with the PCB every year and use Rs. 4,000–5,000 from their RWA funds for this testing.

This pocket was composed of villas and their STP had to be created in an open space, unlike STPS for the apartment buildings, which were in the parking areas. The builder of their villas made the STP in 2009. After the installation, the BWSSB came a few times to inspect it. On one occasion, the BWSSB officers did not know the technology and the estate manager said he had to explain everything. Then they asked him for the design plan for the STP. Another time, the BWSSB officers came when he was on leave. Without his input, the officers recorded that the community was not using their STP. When the manager returned from his leave, he wrote a notice to them, and they came back and asked for photos documenting the functioning of the technology. The manager explained, "They have penalty charges that they calculate on the basis of water usage. So two agencies come to check." The resident added a bit more on the connectivity with centralized systems:

> We treat it and use it [the treated wastewater] for flushing and gardening. We don't feel tempted to just dump it out in the nala nearby [which I observed had a hefty flow of wastewater]. You can raise the funds in the RWA and run everything. Monthly maintenance is Rs. 6,000/month [for each household in his pocket]. It is enough for the STP. The Bangalore Development Area (BBMP) had collected those funds earlier and if it comes under BBMP, they lay the sewer lines. But it is still under the gram panchayat [village governing body] so they use the monthly RWA for their maintenance. No one is laying sewer lines right now. If they lay them for the public drainage system, then we have to connect our lines to it. We use all the water for gardening and flush completely.

Again, the potential for connecting to the centralized systems appeared in this discussion. In their off-grid situation, they perceived the possibility that they could be connected to the centralized systems in the future once their land area was rezoned. This RWA member also perceived a savings in costs because of recycling. Although I imagined that it was still too early to have a full accounting on savings, I took his answer to mean that they expected a savings. It was also possible that this respondent perceived a savings when comparing the recycling costs to the cost of purchasing tanker water, which was set at a much higher rate. A resident in another pocket also explained their decision-making on STP management: "My motivation to use STP water is because it conserves freshwater. I am getting to know the technology and we are planning to maintain [it] ourselves. We have a committee and are discussing whether we will do it ourselves or hand it over to a third party. We are four to five members and will shuffle the responsibilities between us."

REMUNICIPALIZATION AND OPTIMIZATION

Remunicipalization is a term designating the neoliberal process of taking back control of municipal water and sewer services from private companies, a process that has occurred in over thirty-seven countries in the last twenty years.[15] Swings between private and public ownership and control of essential services characterize the history of utilities around the world. The current phase of remunicipalization involves discontent with the profit-motivated private sector, yet the recaptures of public services are a mixed bag. The ideology of remunicipalization rejects total control by private companies, but the neoliberal state may also be deeply engaged with builders, contractors, technology companies and regulators in a profit-based market. The legal institutions may be entrenched in monitoring and the enforcement of regulation. In India this is certainly the case.[16]

The garden STP projects show that public-private partnerships (PPP) and arrangements in the wastewater sector require a guaranteed buyer for the water for them to work. In the drinking water sector, PPP arrangements are built on the guarantee that citizens as water users will pay through a pricing structure set by the authorities. This is the source of contention in the privatization of water around the world, for residents expect the water companies to raise rates and make profits. But in the wastewater field, the question arises: who will be the guaranteed buyer of the treated water to recoup the costs of the project? In the garden STPs, the NDMC has guaranteed to purchase the water for a period of twelve years. In the college and housing society STPs, the institution and the residents must pay out of fluctuating budgets that may change over time. Fluctuating finances may affect the ability of these entities to maintain their systems and consistently produce usable water. Because of the financing source, there is some hedging among housing society members in their thoughts about the future. Will a Cauvery connection end up bailing them out? Will the STP manager be able to make the machines work within the RWA budget? The tentative foundations of some reuse schemes generate insecurity and, in some cases, turn projects into partial reuse rather than the intended full-scale reuse. On the other hand, costs of other sources of water are on the rise and becoming harder for communities to procure.

These cost and availability pressures keep communities in the process of experimenting with reuse machines.

RWAs are at the center of remunicipalization efforts where the state has excused its absence. But these associations are not representative of all citizens. RWAs have been criticized for their exclusionist practices and elite views on world-class cities that push out lower-income residents through sanitizing practices.[17] Since there are no government subsidies for decentralized sanitation projects, RWAs must raise finances for these projects through their monthly dues and use them to hire and pay a salary to a manager, pay for upgrades and repairs of the STPs, and pay operational and maintenance costs.[18] This puts recycled water in the hands of middle- and upper-income communities, and businesses with the capacity to install technologies with their earnings.

Managing microbes appears in many of the comments of the STP managers and residents in this chapter. While it is expected that engineers and consultants have knowledge of the role microbes play in digesting wastewater, residents in housing society projects also have a layperson's knowledge. The awareness of microbes increases as community members engage with their RWAs and with their STP managers. It is possible that at the colleges where STPs are running, the knowledge of microbial activity among water users may be weaker, especially if horticulture is the only application for the treated water. Among housing residents, the knowledge of bacterial activity in the STPs is associated with smells that are experienced when using treated water for toilet flushing. Through the intimacy of toilets, they build an understanding of microbial communities within the context of treatment machines in their building. Earlier hygienic practices of pushing wastage and filth out of sight or into someone else's backyard are now turned upside down.

Builders of apartment complexes jump-start decentralized STP construction. But they are known to skimp on machine components and create less than functional machines. This is where the deception starts. But when RWAs "take over" the STP after a year or two, they are in a position to upgrade the system if they are incentivized by their water scarcity situations.[19] This is the step that was missing in the three cases in chapter 3. Those communities could not effectively inherit projects from government or NGO agencies, whereas in the cases in this chapter the building contract required the turnover to the RWA within a year or two. The decisions are then in the hands of the RWA on how to maintain them. The residents must make choices about technology upgrades using information provided by a network of nongovernmental agencies and then decide how to run the day-to-day operations. Residents make technology choices. Residents and STP managers must also comply with state and national regulations and policies on effluent standards by arranging water quality testing and submitting paperwork. They fill in for the monitoring inadequacies of the state agencies. While dual plumbing is mandated by governments and municipalities, communities and businesses have the freedom to choose technologies and reuses with very little government oversight. In the off-grid examples, the feedback loop for monitoring became tethered to resident reuses as the STP manager answered to the loose requirements of monitoring agencies and consulted with NGOs on repairs and optimization. Tight feedback loops combined with circular reuse loops will appear in the next chapter, in more sophisticated ways.

5 Closed Loops and Emerging Reuse

Now I turn to two kinds of reuses that form the ends of the range of experiments. I describe two categories of reuse, closed loop reuse and emerging reuse. I start with a few of the most advanced technological experiments with closed loop circulation and treatment of wastewater and distinguish them from what is more hopefully referred to as zero discharge. "Zero liquid discharge" is the gold standard in wastewater reuse. However, it is more of a hopeful idiom than a reality. In the closed loop process, the idea is that the gray and black waters will be recycled repeatedly through the user system without being discharged back into the environment. The ideal is zero liquid discharge, and it is the tightest feedback loop for maintaining an effective treatment system. It is important to understand how closed loop wastewater to water reuse systems operate, to offer a more specific view of the requirements, outcomes, and challenges. In describing these closed loop cases, I highlight features and uses that appear in each; they are "dual plumbing," reuse for AC cooling towers, and governance as "tight feedback." In the second half of this chapter, I look at a few emerging reuse cases and highlight the practices of monitoring, the generation of excess water, and raising funds within the community.

ZERO LIQUID DISCHARGE

In wastewater management, zero liquid discharge (ZLD) refers to the condition when there is no treated or untreated wastewater emitted from the water use to reuse loop. Thinking critically about this goal, Isenhour and Reno have noted for other contexts, "Ambitions to completely 'close loops' or reduce waste to 'zero' not only fail to materialize in practice but can serve to conceal forms of excess they continue to dispose of."[1] While I agree that a zero discharge claim can be deceptive, I have found strong support for closed loop systems, especially among communities that experience low water availability. Deceptive practices do occur but are more common in industries and businesses where STPs are not operated as a way to keep costs down. There are other kinds of deceptions that have become more commonplace since the first policies on ZLD were launched in 2015. A short review of this ZLD policy introduces the concept for pollution prevention and water conservation before the respondents' comments on the subject are related through the cases.

At first, the regulatory aim was to get industries to pool their wastewater into "common effluent treatment systems" built in industrial parks or clusters. The aim was to treat the wastewater and then recover valuable ingredients for reuse, such as chromium from leather tanning. The Central Pollution Control Board and the Ministry of EFCC defined zero liquid discharge as: "the installation of facilities and system[s] which will enable industrial effluent for absolute recycling of permeate and converting solute (dissolved organic and inorganic compounds/salts) into residue in the solid form by adopting method of concentration and thermal evaporation. ZLD will be recognized and certified based on two broad parameters, that is, water consumption versus wastewater reused or recycled (permeate) and corresponding solids recovered (percent total dissolved/suspended solids in effluents)."[2]

In 2015, the CPCB issued the first directions for drawing up plans to implement ZLD and use online monitoring to record the quality of effluent emitted from these common effluent systems for textiles, distilleries, and other industries.[3] The National Green Tribunal (NGT) started discussion and enforcement of the policy with their December Judgement in 2015.[4] In this Judgement, the NGT laid out the intentions of the policy put forward by the pollution control boards. The NGT stated:

One of the ways to improve the regulatory regime and to ensure that the industries should adhere to the relevant environmental laws was to enforce ZLD and online monitoring system. In fact, the CPCB [Central Pollution Control Board] had issued directions to the UPPCB [Uttar Pradesh Pollution Control Board] under section 18(1)(b) of the Water Act, 1974 for seeking action plan[s] from industries on implementation of ZLD in identifying industrial sectors in March–April, 2015. It had even issued guidelines for techno-economic feasibility of implementation of ZLD for water polluting industries in June 2015. It required that there shall be compliance with the environmental standards notified under Environment Protection Act of 1986 and to permit the industries to discharge effluents only after compliance. It was acknowledged that ZLD was a necessity and technically exigent. It was also stated that ZLD can be achieved by adopting conventional primary, secondary and tertiary effluent treatment and polishing by filtration and using clean water back into process or domestic use. It also provided an option to select the technical system facilitating achievement of ZLD. In other words, ZLD could be attained by recycling or by achieving no discharge at all by use of appropriate technology.[5]

During the hearings, industry representatives objected to the evolving requirements. Many showed up in the Tribunal and gave their written submissions. The All-India Distillery Association argued that the CPCB and UPPCB had not considered the negative environmental impacts, the burden on natural resources, the economic unviability, the high capital cost, and the long-term sustainability of the directions. The Ministry of Water Resources (MoWR) reissued the rule that there should be "no discharge" of processed wastewater from industry premises, to promote reuse, recycling, and recovery to the fullest extent possible.[6] Some industries followed this policy by spraying their treated wastewater on their land or bio composting the sludge on site. But such a ZLD practice would leave its footprint in the soil and groundwater. After the NGT went further to discuss the practicalities with industries and government agencies, the NGT decided to soften its approach.

> To impose ZLD on such industries would neither be fair nor just. In fact, it will not be in consonance with the requirement of law under relevant Acts. An industry should be permitted to operate, subject to grant of Consent to Operate, by the concerned Board. The CPCB has the competency under law to issue directions under Section 18 of the Water Act. The purpose of empowering Boards with certain powers is to restrict and control pollution. It

is not expected from the Boards to stop the natural growth or restrict industries from operating but compliance to the environmental laws is fundamental to exercise of their powers. The Board must take into consideration . . . the aspects including technology, financial viability, limitations of the unit, [and] process adopted by the industries but in all events ensuring that the discharge of effluents from the unit has to be strictly in compliance with the prescribed standards. No industries, big or small can be permitted to pollute the groundwater, drains, water bodies and environment. To put it simply, the ZLD directives cannot be applied across the board. On the one hand, it would be violative of the rights of the parties while on the other would not be in consonance with the provisions of the relevant environmental acts. ZLD should be applied on [a] case to case basis. The concerned boards should exercise its technical know-how to issue appropriate directions in that behalf. The ultimate purpose is prevention and control of pollution and not an internal management of the plant. Effluent discharge must be strictly within the prescribed norms and the Board in appropriate cases could issue directions with regard to recycle, reuse of the treated effluent appropriately. The ZLD as inferred from the notification dated 7th October, 2016 is incapable of being enforced across the board without reference to the member industries and other relevant aspects afore-stated. Similarly, the Online Monitoring System or Continuous Emission Monitoring System should also be applied on case-to-case basis with reference to the facts and circumstances of the given unit. They must be practicable, for instance, if there is a tannery unit which has consent for processing of hides at a day to be expected to become ZLD or to install Online Monitoring System or Continuous Emission Monitoring System would be opposed to any accepted principles of technology and safeguards of economic advancement. They would be compelled to operate and discharge their effluents only and strictly as per the prescribed norms in default. They would be liable to be shutdown. Another consequential issue that arises in this context . . . [is that] there has to be a specialised, technically sound and dedicated mechanism with every board including CPCB which monitors entire input of [the] Online Monitoring System or Continuous Emission Monitoring System. This monitoring should include not only collection of data but . . . [it should] ensure that actions taken in default and operational deficiencies in the units are rectified within the prescribed time, failing which the unit should be ordered to be closed. The concept of self-regulation would achieve its object, only when there is an effective supervisory control. There have been serious and noticeable drawbacks, deficiencies, and omissions in regulatory regimes else, the current state of industrial clusters, drains, tributaries of the river would not have been prejudicial to such an extent. Continuous calibration by CPCB . . . [must] ensure that the online monitoring system shows the correct values and it must be compared with the actual effluent analysis collected by the Board on regular intervals.[7]

In this long-winded way, the NGT showed its reluctance to regulate industries to the fullest extent of the ZLD policy, and proposed that industries could be regulated loosely, on a "case by case basis." This is the pattern of calculated informality, where rules are invoked and then put aside. By agreeing that ZLD requirements could not be applied "across the Board" to all industries, the NGT put the responsibility back on the pollution control boards to make the final decision. The NGT also imagined that digital monitoring with new sensor devices could eventually bolster online surveillance but explained it would enforce online monitoring on a case-by-case basis. These statements signaled that ZLD would not be uniformly expected for all industries. On the other hand, the focus on online monitoring was eventually applied to five-star hotels. Two of the closed loop cases described in this chapter are situated in five-star hotels.

In housing societies, hotels, and institutions, closed loop wastewater to water systems were presented to our research team as an aspirational goal. Some said, "We are zero discharge," but we were unable to find a real ZLD situation during fieldwork. Some of the off-grid communities in chapter 4 were rather close to closed loop and thus very low discharge when their STPs were functioning properly. In those cases, the costs of operating a closed loop system were a challenge. The costs for STP installation and operation and the production of recycled water are more than the costs for freshwater.

Large hotels and multipurpose housing and office complexes may be able to operate closed loop processes if they have the financial wherewithal to set up an advanced treatment system to produce water usable for AC cooling towers and toilet flushing. In these cases, closed loop systems may save on consumption of freshwater and allow financial savings when freshwater rates are set at higher business and industrial rates. There is a need to look specifically at individual experiments and evaluate their water budgets with and without reuse water.[8]

These advanced treatment experiments are in hotels, universities, housing complexes and residential-office complexes. These represent several of the best examples of closed loop experiments I have seen. The hotels are in Mumbai and Delhi, the university is in Chennai, and the housing complexes are in peri-urban Bangalore and Gurugram (formerly Gurgaon). All these advanced systems were created to address water scarcity. The Mumbai and Delhi hotels needed water for their AC cooling towers and the Mumbai hotel needed water for toilet flushing. The university and its sprawling campus needed water at a time when the city water supply was very short. The housing complexes in Bangalore and Delhi needed water because they were off the piped water grids in the city.

CLOSED-LOOP AND DUAL PLUMBING

In Chennai, engineering professors were instrumental in getting the university administration to invest in a state-of-the-art system to recycle wastewater and reengineer the entire campus for dual plumbing.[9] The university had a population of twenty thousand, with nine thousand students, nine thousand residents, and a floating population of two thousand. The university had a daily need for 2.8 mld of water. The Chennai Metro Water and Sewage Board supplied 1.2–1.8 mld of fresh water, so they bridged the gap with treated wastewater produced on site. During my visit, the campus STP was producing recycled water that was used to flush toilets in the dormitories and provide water for campus gardens, air conditioning, and building and cleaning needs. The STP had been built in a phased manner at a cost of about 20 crores (200,000,000).[10] During the drought of 2015, the campus was able to function with very little city water and they produced their own supply through recycling. A professor in this effort explained to the media:

> The IIT-M STP treats close to 30 lakh litres [3 million litres per day], including 8 lakh litres [800,000 litres per day] of wastewater generated by the IIT-M research park. They send back 8 lakh litres of treated water to the research park for their utilization and 10 lakh litres is routed for flushing and upkeep of greenery. So the remaining 10–12 lakh litres is in excess, which is being diverted into ponds. Once those are full, two groundwater recharge wells are dug up, each having a capacity of 0.5 MLD. This will replenish the fresh water lake. Overall, nothing is wasted.[11]

At IIT-Chennai, water scarcity spurred the development of an on-site recycling unit. Additionally, research and experimentation were integral to the project. Many other universities started following their example. Universities and hotels have also experimented with dual plumbing. The manager of the Engineering Department of the Renaissance Hotel and Conference Center in Mumbai explained that they were first motivated to upgrade and fully utilize their STP after a citywide water shortage affected their ability to run the AC cooling towers for three very large high-rise hotel buildings. At the time, they were using a smaller STP and reusing the treated water for horticulture. To upgrade the facility, they created two STPs with capacities to treat 750 kld and 650 kld, and placed them under the backyard of the hotel complex. From these two STPs they were able to produce enough water for their AC cooling towers, flush toilets, and gardens on the property. The management installed dual (also called double) plumbing in the entire hotel to direct the recycled water to the specific purpose of flushing toilets. The piped water provided by the city was used for the kitchen, faucets, and showers in the rooms. As the manager explained, the problem with older hotels was that they did not have separate plumbing. It was hard to re-plumb and lay pipelines without shutting down the hotel or disturbing the guests. The manager explained:

> The challenge is that the old hotels do not have the separate piping system for flushing. Most have one pipe system for all uses in bathroom. The properties that have separate line for flushing can easily do it. A few old hotels have it but not all. Retro fit is expensive and it's hard to do in a hotel where we cannot disturb guests and can't shut down because we will lose revenue. But now all new hotels and malls are designing for separate lines. Before water was readily available in India, but now it is not available. It is not the case now that we have all the water we need. The water table is going down to 250 feet or more. It is too much. It used to be 20–30 feet. So that is why everyone is thinking about how to save the water. So they are thinking of recycling, but we need to reduce our use also. So we are going for low-flow aerators for wash basins, 4.5 liters per minute plus dual flushing system. We have completed one building for low flow in the wash basin. It is 10.8 liters per minute in the other buildings.[12]

He told me they see a direct savings in their water bills:

> The regulation is one thing. It is our responsibility to give back to society. We can afford to have more fresh water, but if we do, that means someone is getting less. It is affecting our overall bottom line. In 2015, we were having a problem from [our] STP and were not getting enough water. We went for

Figure 13. Treated water from an advanced system in a five-star hotel. (Photo by Kelly D. Alley)

upgradation—that time using for gardening and a little bit for cooling tower. We were using a little for flushing but mixing it with fresh water in the tank before going to the rooms. Then we did modification and we have reduced intake consumption by 45 percent. We get Rs. 110 per kiloliter. That is our cost. So if I am getting 400 kiloliters from the STP every day, I am saving. Plus, you get a good feeling that you are doing your bit. It doesn't go under CSR because it is a requirement.

The manager mentioned that it cost them Rs. 110 per kiloliter to produce treated and reusable wastewater. This is a very high price for a kiloliter of water when compared to other sources available to communities in housing societies and residents across the country. But among business and industrial rates for water, this price point was more attractive, especially for five-star hotels. On government price sheets, the cost per kiloliter at the high volumetric rate was Rs. 140 or more.[13] Comparing this price to the price paid by a low-income household, where the rates are Rs. 7 to 14 or to the high-rise apartment dwellers where the highest volumetric rate is Rs. 46, the corporate rate is substantially higher.

CLOSED LOOP FOR AC COOLING TOWERS

This five-star hotel in Mumbai was using treated wastewater for its AC cooling towers. Large institutional water users are turning to recycled water for this use. Regulators are also requiring large businesses such as five-star deluxe hotels to use this water for cooling towers. In 2017, the NGT appointed a team to monitor eight five-star deluxe hotels and imposed fines on those that needed to upgrade their systems. In upgrading, many hotel managers engineered new systems to reuse their treated water for AC cooling towers.[14] In one five-star deluxe hotel in Delhi, I spoke with the staff engineer who asked that we not name his establishment. He provided a good deal of information that I use to think further on the costs to produce reusable water in five-star hotels. By 2018, they had installed a 450 kld STP to treat their kitchen and bathroom wastewater and reuse that water for their AC cooling towers, for gardening and for flush toilets in the staff quarters. I asked if they were using the reuse water for toilet flushing in the guest bathrooms. "Not for guests," he replied, "because with STP water after prolonged uses the toilet bowl becomes yellow and green. We are trying to improve our quality so you might see it in the future." They upgraded their STP in 2017 and installed a separate effluent treatment plant (ETP) to treat their wastewater from laundry. "We can't use the STP [the original one] because it [the laundry wastewater] has chemicals which kill the bacteria." They were planning to reuse the treated water from the laundry but encountered a problem with the TDS. "In higher quantity you get white film on the laundry. We stopped reusing for that reason. We can use ETP water for cooling towers or garden," he explained. We went on to discuss the ZLD policy.

KA: Do you ever discharge into the sewer when you can't use it after treating it?

MANAGER: We try not to discharge as we are supposed to be a zero-discharge hotel. So we don't do that.

KA: Did you decide to be ZLD or was it from the court or NGT?

MANAGER: It is our own initiative to be a zero discharge hotel and we are trying to maintain it.

KA: So when did you declare that you were ZLD?

MANAGER: Not sure. I wasn't present in hotel then. But we are trying to be ZLD.

NM: Still you are not 100 percent?

MANAGER: We are trying, but there is always some water that you have to drain out. There is backwash water. Because you have to clean your filter and it is pure calcium or sludge. We tend to throw it outside.

KA: How many of the other XXX hotels are trying to be zero discharge?

MANAGER: Not sure, Ma'am. We have STPs and reuse everywhere.

The ZLD possibility sounded promising in these accounts of the closed loop process. In this interview, I was able to obtain their water costs and verify the higher rates that are assigned to a five-star hotel for water provided by the NDMC. They bought 250 kl of piped water from the NDMC every day and paid Rs. 225 per kl for that water. That commercial rate is much higher than the lowest residential rate bracket of Rs. 7 per kl and higher than the rate paid by the five-star hotel in Mumbai. Additionally, they were purchasing 100–150 kl of tanker water every day from private tankers authorized by the NDMC and paying Rs. 115–120 per kl. Those tankers were authorized by the NDMC to extract water from borewells around the city. They paid a sewer charge to the NDMC as a percentage of the piped water they received. The engineer estimated that the hotel's cost to run their STP and produce reuse water was around Rs. 30–35 per kl. This estimated cost included the operations, human power, chemicals, and repairs involved with the STP, but it did not include the original purchase and installation of the machines and parts. The manager put the installation cost at Rs. 1.5 crores (15,000,000). He continued, "So this is why we are reusing water for cooling towers. The initial cost is very high for the STP since there will be a smell. So we have to have an advanced system. They produce harmful gases and need ventilation and then having space and OM costs. Against all these costs, we don't save much. But it is a definite advantage." I will return to these estimates on costs in chapter 7.

At the time of my visit, the engineer told me that the NGT and the Delhi Pollution Control Committee (DPCC) were monitoring their STP functions from time to time. The hotel's engineer was submitting water quality reports while the NGT and DPCC were "dropping in to check," and taking samples and readings. A few months before, the DPCC asked the hotel to install an online monitoring system. The engineer added, "We have installed it in our filter water tank and the data is uploaded every five minutes to DPCC so we know the quality. Now it is regulated nicely at the outlet of the STP water. The link is available to them, and we have the ID password. Otherwise, they show up randomly." It was not clear if this very loose monitoring loop was sufficient for ensuring water quality standards, but the hotel's need to produce water of a certain quality worked as a form of self-monitoring. I will return to this point later.

CLOSED LOOP IN OFF-GRID HOUSING

To look at a closed loop system in housing, I return to Bangalore and highlight one apartment complex that stands as an exemplary example of a closed loop system in a housing complex. This community claimed to use 40 kl of recycled water and 50 kl of fresh water every day. They had ninety occupied flats in the complex. They had several kinds of water supplied to them: recycled wastewater for flushing, borewell water for non-potable uses, rainwater that was added to the non-potable supply, and tanker water for potable and non-potable needs. Two borewells gave them 50 kl of groundwater every day and private tankers provided a bit extra during a month of the summer. They said that if they had good rains then their borewells would sustain them through the summer. They put 50 kl of borewell water into their underground tank every day and mixed it with the harvested rainwater that was directed into this underground tank. When that tank filled, the excess water was directed into the mouth of the borewell to recharge the groundwater. The water in the underground tank was then pumped up to the individual overhead tanks for each apartment. With good rain, their two borewells could be recharged. When they had

Figure 14. The STP for an upper-income housing community working toward a closed loop process. (Photo by Kelly D. Alley)

excess water in the rainy season, they overwatered the gardens to recharge the groundwater. They were completely off-grid and received no piped water from the city. Their own STP treated their wastewater through the secondary stage, and they sent the sludge away in private tankers. They got their desludging done every month or two and paid for the contract work from their RWA maintenance fees.

After hearing the details, I asked a leader of the RWA, "You are fully closed loop, you don't let anything out?" "Yes," she replied. 'We are zero waste. Only thing we are sending out is sanitary and kitchen waste. Dry waste is going out, paper and plastics. The garden is completely STP water. We have sprinklers and pipes." I asked about whether they had any restrictions for children playing in the garden space. There were no restrictions, but she said parents were aware that the garden was watered with STP water. The BWSSB did not regulate the quality of the reuse water.

Figure 15. Treated water used for gardening in an upper-income community working toward a closed loop process. (Photo by Kelly D. Alley)

They experienced the problem of odor when using the STP water for flushing. One resident remarked,

> The flush used to stink, and the stink would come into the drawing room. But they added some filters, and now the stink is gone. My house is near the STP, and in the morning it used to smell when they started it. But now they have cleaned the filters and added something, and it is better. Our flush tanks get sludge in the bottom and can create fungus. So, we have to get our tanks cleaned. We used to clean bimonthly, but now that the water is better we have to clean the toilet tank less. The grass would be yellow at first. Now this water is good for the garden.

Another resident added:

> The STP requires nonstop attention. It has to be taken care of always. It has to be treated with a lot of care. There have been moments when it has not worked and it got smelly. And the water is used in flushing and gardening.

There have been complaints that it is smelling and the water quality is not good. They have put filters [in], so for the last four or five months the water quality has improved. Last year it was bad, but we are a society and we are very cautious. We were motivating people and saying please bear with us. Please use treated water because the groundwater level has really been going down and there is hardly any freshwater left. There is crisis. So all the residents switched over to freshwater because our STP water was smelly. But now I think they have switched back. So, our STP is doing well.

This housing society had a strong governance board and was active daily in working out water and wastewater needs and problems. They hired a full-time maintenance person to look after the STP and the system appeared to function well. They were self-monitoring the treated effluent to make sure their reuse water was within safe standards. The governance, being fully decentralized, gave the community the freedom to decide to operate within a closed loop, as they adjusted each part of the loop when optimization was needed. Optimization was not only a machine function but part of the management of groundwater and rainwater. Their practices accounted for wider hydrosocial cycles, as they directed rainwater and treated wastewater flows to the groundwater for recharge and then benefited from that recharge in their extractions around the year.

CLOSED LOOP AND SELF-MONITORING

Many of the new upscale residential and office buildings in peri-urban Bangalore, Chennai, and Delhi have advanced STPs to produce water for gardening, toilet flushing, and AC cooling towers. These advanced STPs may run on activated sludge bioreactors (where bacteria are introduced to an aeration tank to spur multiplication), MBRs (membrane bioreactors), MBBRs (moving bed bioreactors), or sequential batch reactors (SBRs) combined with tertiary filters such as sand, carbon, UV radiation, ozone, and RO membranes. The installation and operational costs for these bioreactors and filtration systems are much higher than the SBT, phytorid and MBR STPs introduced in previous cases.

In residential and office buildings with advanced systems, the operations are not managed directly by residents, but are done by the builder in conjunction with a management company. In towns across the Yamuna River within the National Capital Region of Delhi but in the state of Uttar Pradesh, high-rise apartment buildings stretch across the landscape for as far as the eye can see. The high-rise apartment buildings closer to the banks of the Yamuna River were completely dependent on groundwater for their water supply, and this groundwater was recharged from the monsoon flows in the river. It appeared that riverbank residents were using groundwater without government restrictions and were less interested in running their STPs properly or producing reusable water.[15]

My research colleagues and I found one housing community that was an outlier, in the sense that the builder was dedicated to a closed loop approach from the beginning of construction. When we entered the underground room where the STP for this complex was located, we were assaulted by the loud sounds of the machinery for the 425 kld STP. The STP manager explained that they were continuously moving the wastewater out of the storage chamber after treatment. It ran for eighteen hours a day. They used an MBBR technology that had media in the aeration tanks to improve digestion. We had a short conversation about what media do in the digestion tank:

ENGINEER: Our bacteria need surface to build the colonies. Waste comes and is their food and generates bacteria.

KA: So bacteria sticks on the media?

ENGINEER: Bacteria need surface to increase. So they use the media for surface.

KA: The media is like plastic stuff.

ENGINEER: The waste that comes, the bacteria eats it.

KA: After eating what happens to the bacteria?

ENGINEER: We have to watch the amount of bacteria. If it grows too much, we remove some. Otherwise, it won't work. We do sludge checks. We keep it [the bacteria] to 300 per ml. If it goes up to 500 or more, we remove some. We give air in this tank. It is aerobic. Then this is the settling tank, and the sludge settles in this, and the clear water goes into another tank. Then there is the chlorine contact tank.

A loud clanking sound was blurring our words. He continued:

ENGINEER: Then after we use this for flushing and gardening. We use 100 percent of this STP water. We are zero liquid discharge. Here is the final outlet.

KA: There is no smell because you add chlorine?

ENGINEER: Yes, no smell. We do weekly tests. But BOD is maximum less than 10. XXX lab is here, and we use that. Our PCB allows up to 30 BOD, but for flushing we need 10 BOD. TSS [total suspended solids] is 4–7. We almost remove it. DO [dissolved oxygen] test we do, and COD [chemical oxygen demand] test and TSS. We have a carbon filter also. I will show you. The RWA doesn't handle this. XXX agency does it for the building. The contract is for one year. We use [a] double media filter with gravel and sand layers. If the TSS gums up, we backwash it to clear out the sludge from the gravel. We can use the sludge for fertilizer. We use it for the garden. This is our final storage tank. The builder built the dual plumbing because it is compulsory. Without this plan, you cannot get approval. This building is five years old. Around 2012–13, it became compulsory. Before that, it was optional. But then the NGT made it clearly compulsory. We monitor ourselves. PCB and NGT come sometimes. It is possible they will do surprise visits. We use disinfection for flushing water—some extra tanks are for that.

KA: What exactly is that? Is it ozone gas in there? For sterilization?

ENGINEER: It kills all the pathogens. Chlorine can't kill all the bacteria and pathogens, but ozone is more thorough. We have an oxygen concentrator but it's not on yet. We will provide oxygen as needed. This is the final stage. FCC [fecal coliform count] at the outlet is less than 100. One flat uses 150 kl per person in peak hours and 120 kl per person per day. It is their second year with this STP. The building is four years old and the builder did it for 2 years then we took over. We have five men working here on different shifts. The building pays XXX to maintain. We give them a monthly costing. We clean as well.

Engineers monitor the wastewater using several parameters of water quality. Bacterial growth has to be managed, and then bacterial communities have to be killed off with chlorine at the tertiary stage. This work requires more biochemical expertise, so the STP was run by a facilities management firm hired on a one-year contract to handle all operation and maintenance tasks. The financing came from the RWA fees that residents paid monthly. As the managing engineer told us, the firm monitors the effluent quality and has the incentive to achieve higher standards for reuse.

We found practices of self-monitoring in another high-rise office-cum-residential complex in Gurugram, the arid southwestern corner of the National Capital Region of Delhi in the state of Haryana. As in the

Figure 16. An MBBR bioreactor underground in a residential-office complex. (Photo by Kelly D. Alley)

Figure 17. Media in the MBBR bioreactor. (Photo by Kelly D. Alley)

Figure 18. Media pieces for the bioreactor. (Photo by Kelly D. Alley)

southeastern and southwestern parts of the NCR, water scarcity was a
strong driver of the interest in reusing wastewater. At the time of our visit,
the large building complex was in the process of construction. The office
section had been completed and the residential sections were under construction. The STP in this complex was made to treat 425 kl of the wastewater generated from the office and residential sections of the building
complex every day. The water supplies for the building came from the Haryana Urban Development Authority and from private tankers. There were
older borewells still working in the area, but new borings were banned by
the Central Ground Water Authority.[16] Since this office-resident complex
was new, they were not able to procure a borewell permit. In this predic-
ament, the builder had to make sure that the reuse of wastewater could
be maximized. This maximization also extended to construction projects
nearby, where construction companies were required to use treated wastewater and not surface or groundwater for construction. The builder of this
complex had used treated water from a prefab MBBR STP and from a
centralized STP operated by the Haryana state government.[17] The builder

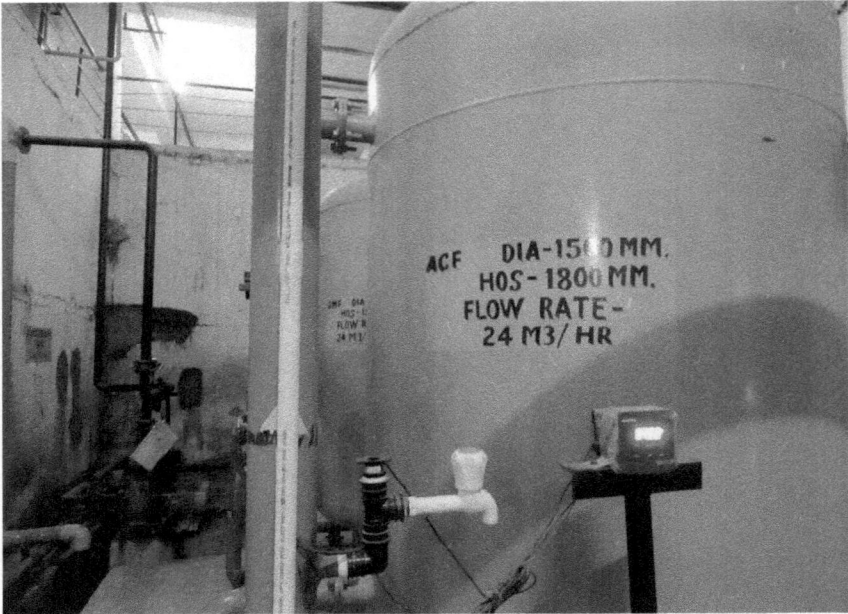

Figure 19. Dual media and activated carbon filtration devices phase in an office-residential complex. (Photo by Kelly D. Alley)

made the design for the STP, chose the technology and the location, and supervised the construction of the facility.

The building authority hired a management company to carry out operations and follow the maintenance and monitoring requirements set down by the state government pollution control board. The management company said the PCB (Pollution Control Board) was not monitoring the standard of the reusable water but rather "suggesting" what they should do. He explained, "There is a manual. Engineers give guidance and the government provides the parameters on zero discharge." The facilities management company had an agreement with the builder to operate, maintain and monitor the STP for five years.

At the time of our visit, the management company had been working with the STP for six months. The STP staff worked around the clock in three shifts. Each manager earned around Rs. 16,000–17,000 in salary per month. In this complex, there were two storage tanks, one for hard water and one for soft water. The soft water was used for the AC cooling towers. The manager of the STP explained that they needed to keep the water circulating through the building to prevent water loss. Wastewater reuse was expected to reduce freshwater use by 60 percent. The demand for air conditioning in commercial office buildings pushes up water demands well beyond what housing societies use for toilet flushing.

EMERGING REUSE AND EXCESS WATER

To introduce the other end of the spectrum, I conclude with a few examples of emerging reuse cases. Emerging reuse experiments are open-ended and evolving. Communities are trying to figure out what to do, how to confront their wastewater problems, how to think about solutions and how to operationalize them. They are grappling with emerging conditions without infrastructural aids such as dual plumbing systems. However, most have a rudimentary STP and can build and upgrade for reuse options. It is important to look at emerging use cases to understand the challenges they face in the middle ground between treatment and reuse.

In a housing society in peri-urban Chennai, the RWA had just started developing its STP. They inherited it from the contractor who had installed the unit during construction. At the time of my visit, they had a MBBR or moving bed bioreactor built underground on the side of the ground-level parking deck. The apartments were built above the parking deck. The tanks for collecting the wastewater and aerating it were accessible via square manholes lodged in the deck. There was also a room under the parking deck where the sand and carbon filters were situated. After running through the MBBR bioreactor, the water went through these sand and carbon filters. The builder had installed monitoring equipment so that the residents could determine the real-time quality of the treated water. This was to aid in reuse of the water for toilet flushing. However, the RWA started to realize the true operational costs when they took over the unit from the builder after the first year. After taking over the facility, the RWA decided to remove the monitoring equipment to save on their operational costs. Because they did not have a dual plumbing system set up and were not reusing the water, they did

Figure 20. An MBR system on the side of the parking deck of a middle-income housing society. (Photo by Kelly D. Alley)

not see the need to use real-time monitoring instruments. The RWA selected a concession company and paid for staff to maintain the STP and repairs and supplies.

The maintenance company did not mind me asking questions, but they did not want me to take any water samples, which I was not qualified to do anyway. My guide, a PhD student at a nearby university, and the onsite maintenance manager told me that the community was using the treated water for gardening. Since their gardens were small, they were generating excess reuse water. I asked about what they were doing with their excess treated water and the maintenance manager explained that they had to pay for a tanker to come every day and take it away. They could not store the water, so they needed to send it out every day. I was able to see the tanker that they used. The manager was not sure where the tanker water went, perhaps to a municipal STP or to the gardens of Chennai. It was not clear to them or to me. "The tanker comes every day," he and the guide said. "Once they install the UV system and use it for toilet flushing, they will be able to reuse the water fully." The temporary system seemed wasteful to me, since the RWA had to pay for the STP maintenance and pay to have the treated water hauled off as well. But I could understand that it was a way to patch the disconnected system in the immediate term.[18]

In their off-grid location, they had no piped water from the municipality, so they were buying tanker water for household uses and for drinking. They had borewell water, but it was not usable for potable or human contact purposes. They had four overhead tanks—one for metro tanker water (for drinking), which they filled every day; one for borewell water, which was salty and only usable for washing household items; one for borewell water used for toilet flushing; and one for borewell water for firefighting. The guides told me that there was "no social acceptance" for using the treated wastewater for washing purposes. It is "the mindset" of those who "don't trust the [municipal] water." The community was not experimenting with reuse, and so the acceptability, as reported by the STP manager, was low.

EMERGING REUSE AND GATHERING FUNDS

In another peri-urban area of Bangalore busy with IT corporate buildings, residents were paying for tankers during the summer when water was scarce, when their borewells were not producing water, or when demand exceeded supply. There, residents were paying Rs. 400 for five-thousand-liter tanker of water. In the peak summer months, the tanker prices could increase from Rs. 500 to 700. As one RWA member explained, "You have to negotiate. Seven to eight tankers per day in summer so [Rs.] 3,000 per day. In one month, Rs. 90,000 to 1 lakh [100,000] just for the water. In summer, the water level goes down, so the borewell might not reach the water. It is 950 feet deep. We have two borewells here. The water is costly, and we can save money by using the STP water for the garden.

Initially, the builder was managing their STP, but that contract had expired by the time of our team visit. So, they hired someone to look after operation and maintenance. The builder installed what they called, "old, cheaper technology." When they began managing the facility themselves, they decided they wanted "the latest ecofriendly technology that could operate without electricity." They contacted Ecoparadigm, and got a baffled reactor installed with their guidance. One RWA resident leader explained about the decision-making process:

> Gathering the money was not the issue. Selecting the right tech and making people convinced of the technology was an issue. The builder had planned an STP with a cost of 50 lakh rupees (5,000,000) with old technology. [It was] a sequential batch reactor with full-time electricity. But we wanted low maintenance. That is when we met Ecoparadigm and thought it was a good technology and required no electricity. No maintenance for the first two years and then Rs. 40,000 for desludging and other maintenance after that. Nothing should protrude out or smell. We didn't have a place, so we got it under the ground and can walk above it. People are concerned about the cost for dual plumbing. We have already paid for the installation. Acceptance is there for flushing. [We have] 130 households. Fifteen to twenty percent of residents have good education. If you use[d] STP water in flush, you would get [a] stinking smell. But when you talk about the water scarcity in the future, no religious issues will say no to that. If we are able to convince them about PGF [planted gravel filter] or whatever, that concern will go.

I asked about how they selected their consultant. Another RWA member explained that they did a lot of googling and people came and explained their technology. Some companies proposed a cost of 65 lakh (6,500,000) rupees requiring 24/7 staff. He commented,

> We didn't want that. We now have no manual observation. We have a pump to take the water to a drain. We had a discussion with the lake authorities, and we say we will do testing for PCB [Pollution Control Board]. Sixty thousand liters generating daily, and we can reuse 30k for toilet and 10k for garden. So how to dispose of the rest?[19] The government should help us, as we don't have space to use the remaining water. Right now it goes to the raja kaluve [an erstwhile stormwater drain] after treatment. All are sending their raw sewage there. We are at least sending the treated water there. It is two hundred meters away. We laid our pipe until there, and need a pump just to dispose it over there. We pay property tax. Why should we pay sewage charge when there is no proper drainage? Our maintenance includes water, electricity, security, housekeeping, amenities. We were running on

Figure 21. Pipes connecting the apartment building's sewer system to the planted filter. (Photo by Kelly D. Alley)

soak pits before we had our STPs. We were depending on sewage tenders [for desludging] and that shot up our maintenance heavily. So that cost helped us to convince the society to get the STP. Problem with sand and carbon filters is that they are above the ground. But then we would need a motor. But we want zero maintenance and eco-friendly. So we can use a combination.

I asked if they thought they would be able to save money by reusing their treated water. He thought they could, given the information he could obtain from googling. He continued,

We know that we can reuse 30 percent of the water and should be able to save 30 percent on the expenses. Especially during the summer when we spend around Rs. 6,000 per day for tanker water. Eight to ten tankers we buy daily in the summer for 5,000 rupees per day. Most of the water—about 30 percent—goes into flushing. So if we are able to reuse that water, pumping the water to an overhead tank would not be that much. We need a 2 hp pump and it would not generate so much of a bill for one hour each day. We have two pumps for our borewell but one dried up two days back. It goes down 1,000 ft. We are hoping with this rain, the level will come back.

The secretary of the housing society then entered the discussion and added that he liked the idea of reuse. He said, "This [their STP] is not totally maintenance free but nearly maintenance free. Rs. 40,000 every two years for desludging and reintroducing bacteria is reasonable. People are interested to know when we can break even. For the 50 lakhs [5,000,000] we have spent, we want to see if there is a return on our investment. We have an idea to do drip irrigation on gardens around the periphery. Ecoparadigm said we can use this water for gardening without chlorine. So do we need to put chlorine in the tank itself?"

They had questions I could not answer. We continued to discuss the other STPs and models we had seen. A member of our research team then asked whether they thought their STP made a larger impact on society and one member answered: "In this area, our apartment is one of the biggest apartments. There are a minimum of twenty apartment buildings on this road. Some are under construction or already constructed. We have a street group. Some people already know that people around us are implementing the STP. Some contacted us and got the contact of Ecoparadigm and discussed with them. So they tell their RWA. So some are eager to know about these eco-technologies toward STPs." He added that many of the RWA board members had the experience of using recycled wastewater in their IT offices where they reuse wastewater for toilet flushing. "All the IT companies are using recycled water for flushing. So we don't have any issues." We asked if it was of good quality. "Yes, I mean there is no smell. They have a separate person taking care of the restroom. You don't get the feeling it is treated water. But we are looking at a future where we will be fighting each other for water."

Doing the math for this building, they calculated that they consumed 70 kl of water per day on average and generated 60 kl of wastewater. Their cost for upgrading the STP was 55 lakh rupees (5,500,000), which was an add-on to what the builder had paid to set up a simple system. They found out that dual plumbing would cost Rs. 15,000 per flat. They were also interested in creating fishponds with the treated water. One board member added, "To reuse it [the treated water] would be more expensive, but we could make a channel around the complex with ponds to grow fish. And also use for flush and get revenue for growing lotus and fish. We can make handrails to keep the children out. In lower level, we have fish and top layer we grow other fish. Is it possible to do that with this water? Could the fish be consumed by people? Would there be health issues?"

Again, I felt helpless at not being able to provide studied answers to their questions. They had heard of this possibility from the consultancy. They were looking for ideas and answers to their questions. They were exploring options. Their questions underscored the central role of the RWA and its key members in making critical decisions and funding their decisions in-house. The fact that they were envisioning fish cultivation also showed that communities could create new options that might generate other sources of food or income.

MONITORING AS FEEDBACK IN CYCLES AND LOOPS

Closed loop is a way of conceptualizing a circular system instead of a linear one. The goal is to make the most of water resources and generate as little wastewater as possible. Cycles are part of the transformations of water, as well as carbon, energy, and waste. Feedback is important in loops and cycles, such as the closed loop experiments and the broader hydrosocial cycle. Feedbacks can generate sustainable self-regulatory processes when a group of people aim to produce reusable water. But water quality within feedback loops changes over time, and it cannot be replicated to the same standards every time. Monitoring agencies and individuals work at different times and at different speeds to generate the feedbacks of information needed to optimize machines, enhance microbial digestions, and adjust water qualities. The broader hydrological and social cycles are constantly changing with many other uses and climatic changes. Feedback instruments are embedded in STP machines to guide processes of treatment and help to alert operators to malfunctions. Microbes are living communities that must be measured and optimized, making water quality testing an essential feedback instrument in the digestion of wastewater. All these feedbacks work in a looped rather than linear fashion.

In the three closed loop projects described in this chapter, water circulated on a limited path from treatment to on-site reuse and then back into treatment. The businesses and institutions handled the responsibilities of operating and maintaining the feedback processes. In the peri-urban areas of the National Capital Region of Delhi, management companies were engaged in operating, maintaining, and monitoring the STPs in residential and office buildings. This created a nexus between a builder, a management company, and an RWA or between an RWA and a management company. The arrangement between an RWA and a management company was more common in our survey work and RWAs were generally handed the STPs by builders after one or two years. The management companies hired by RWAs were responsive to the rules and water quality standards enforced by the PCBs and municipalities and sent their samples to authorized PCB labs for testing.

Reuse options motivate users to monitor the quality of the wastewater using the government's standards and parameters, which change over time. In universities, five-star hotels, and large residential and residential-office buildings, self-monitoring ensures that the reuse water is acceptable for AC cooling towers and toilet flushing. Online monitoring systems were being installed in universities and five-star hotels but were not used by RWAs. On the IIT-M campus, engineering professors and their students were involved in monitoring machine performance to forward their research interests. These new monitoring arrangements were transforming consumers into self-interested operators and monitors. Monitoring was creating tight feedback loops within larger hydro-social cycles.

In the tight feedback loops within housing societies, groundwater was pumped up for on-site consumption and then turned into wastewater. It was treated and then reused for horticulture. The reuse and harvesting of rainwater recharged groundwater. In these tighter feedback loops, residents could recognize the waters as usable or unusable for different purposes. The closed loops gave communities more control over their water supply. Decentralized governance was shaped by choices on technologies, the implementation and use of dual plumbing, the upgrading of technologies to produce reuse water, community financing, and monitoring of water quality. While some loops discharged very little and came close to a ZLD status, there was always water entering and exiting the loop. In emerging cases, loops were not yet formed, and water was diffused into wider hydrosocial cycles. I will return to this conceptualization of loops and cycles in the concluding chapter.

6 Pretend Machines

The previous two chapters have presented inspirational cases of wastewater reuse that will serve as models for experiments in India and other countries. As communities and businesses face water stresses, they will continue to face challenges. In this chapter, I zoom in on hotels and ashrams, or places of religious learning and residence, in the town of Rishikesh in the state of Uttarakhand, to highlight a few more of the contradictory and deceptive human practices that complicate the goal of building and running treatment machines. It is important to look more deeply at deception because it is the underside of failure. As the hidden transcript, deceptions must be revealed in the experimentation process to avoid false conclusions. Examining deceptive practices can help to understand the challenges of compliance and monitoring exercises required by the courts and the NGT.

Everyday practices produce and expend symbolic, economic, and technological capital. As the production of paperwork grows more important in the construction and monitoring of small-scale treatment machines, private businesses and housing society groups are becoming more adept at noncompliance with documented rules.[1] This is a critical problem not only for the state, whose centralized environmental monitoring agencies (mainly the CPCB and PCBs) are already understaffed and underfinanced, but it is also a problem for municipalities and for residents across the urban spectrum.[2] As practices of noncompliance become more complex, the researcher has to examine how businesses and communities respond to central legal and administrative orders and policies over time. This could not be done in the previous case studies because return trips to most of the sites were not possible. But in this chapter, I show how it is possible in one town where I returned over a period of six years. During this time, I tried to understand responses to court and NGT orders and government policies in terms of the real, performative, or deceptive functioning of treatment machines.

Compliance and the elimination of corruption are important to the delivery of public goods and services such as water and sanitation.[3] In this book, I have read corruption not so much from the economic and monetary activities involved with installing and operating sewage treatment machines and building sanitation infrastructure as from the facades of functionality, which are far more telling of the extent to which investments have been properly utilized. In other words, the machines and infrastructure may be there in all their parts and pieces but as my previous chapters have shown, they may not work!

Attending to fake machines and structured practices of pretending compliance raised questions about discourse and normativity. I had to decide whether the truth statements or the lies were normative facts. I kept thinking that respondents would tell me the truth about their machines— I wanted to know the truth—but I found that lying and fibbing were appearing so common among Rishikesh businesses as to be the de facto normative order. This normative situation led me to consider the practice of noncompliance as an everyday practice of intentional misconduct in addressing the challenges of maintaining sanitation machines and city infrastructure.[4]

For this discussion, I gain insight from another paper on decentralized treatment plants written by Starkl et al.[5] In an analysis of fifty-eight projects in India, they elaborated their definitions of success and failure, to note that hidden failures may exist behind the appearance of functioning systems. They added that layers of known and hidden failure are part of the complexities of decentralized systems where the enforcement and monitoring is not spread across all facilities evenly. Their qualification on success is quite fitting here:

> For this paper, "apparent success" is success in the view of local experts (step 1 of data collection) and "actual success" is success in the view of a closer inspection by international expert teams (step 2). "Hidden failure" is apparent success that is not actual success. This paper emphasizes hidden failures, as the subsample of hidden failures appears to be random, justifying statistical methodology. (In step 1, data could not be collected at random, as insight by the local experts about the systems was needed. However, as hidden failures result from a lack of information, one could not intentionally search for them.) "Hidden success" was not observed.[6]

As Starkl's group and others have noted, the ingredients of success and failure are rather murky. In many of the machines in the housing and hotel complexes I have reviewed so far, public-private arrangements were shaped according to the specific inputs and responsibilities of NGOs, residents, and business managers, and they all appeared to engage in acceptance of technologies and the purpose of treating wastewater. This means that they were agreeing to follow practices deemed beneficial for the public good and for water conservation. Likewise, the cases in this chapter demonstrate the willingness of business owners to install machines for decentralized treatment. But after probing their practices through several different kinds of investigation, I found a general unwillingness to run these machines effectively. In their dispositions as machines that appeared to work, they began to look more like deceptions and fake infrastructure. At some point in my field research, I had to stop and remind myself that the inductive method meant I needed to open my eyes and accept the reality. These machines were really not working!

I collected data in the town of Rishikesh over a period of six years during several visits. Through my own efforts and in concert with research colleagues and student research teams, I developed more than twenty key informants and conducted ten focus groups and forty interviews with owners and managers of large, medium, and small hotels. A student research team and I administered our survey to sixty-six small and large hotels and managers of ashrams. I also collected official and company records and visited the National Green Tribunal in Delhi (which covers this region) and the High Court of Uttarakhand as a participant observer. In the Tribunal and the Uttarakhand High Court, I met advocates and legal practitioners and discussed the histories and previous actions of the relevant legal cases. I exchanged perspectives and shared data with the university research teams that were monitoring the centralized STPs in this state and toured the centralized STPs in Rishikesh with professors and members of the state engineering agency, the Pey Jal Nigam.

RISHIKESH

The town of Rishikesh is a sacred city known worldwide for its pilgrimage culture and religiously motivated tourists. The town also falls under the purview of the 2015 National Green Tribunal orders that direct all housing societies and five-star hotels to install STPs. The NGT also ordered smaller hotels and ashrams, places of Hindu religious learning and practice, to tender, install, and operate their own STPs on their premises and treat their wastewater before discharging it into the city drains. In Rishikesh, I found that responses to this order were uneven and variable and were shaped by a business's ownership structure, location in the town's centralized infrastructure, geographical characteristics such as proximity to the river or a groundwater source, STP installation and operational costs, and tax structure. Hotels pay an income and goods and services (GST) tax, but ashrams do not. I noticed that this difference in tax requirements ruffled the feathers of a few hotel managers and undermined their interest in complying with environmental regulations.

In Rishikesh, I focused the inquiry in two main areas of pilgrim and tourist activity, the Nagar Panchayat of Swargashram-Jonk and the Gram Panchayat of Tapovan. These two neighborhoods provide an interesting internal contrast while posing a larger set of problems related to uneven and performative practices. Across the river from the water-stressed neighborhood of Tapovan, just north of the center of Rishikesh, is the water-rich neighborhood of Swargashram.

During the twentieth century, the area known today as Swargashram was established as a resting point en route into the Himalayas, by way of the Char Dham Yatra. The Char Dham Yatra (the pilgrimage to four holy places) links the sacred places of Yamunotri, Gangotri, Kedarnath, and

Badrinath. In the late 1890s, the sadhu (Hindu mendicant) Kali Kamli-wala settled in Swargashram and started giving water, food, and shelter to visitors. Gradually Kali Kamliwala's retreat grew and passed after his death to Swami Atma Prakash. Atma Prakash and other sadhus established the Swargashram Trust in 1938 to govern the land area of Swargashram extending up to Neelkanth. Over time, other ashrams were established. Geeta Bhawan was built in 1939 and Parmarth Niketan was created in 1942 by the Swami Sukhdevanand Trust, in memory of the life of Swami Sukhdevanand.

The Swargashram Trust remains the largest in terms of land today, retaining around 40 percent of the current area of Nagar Panchayat, Swargashram-Jonk. Over time, the Swargashram Trust donated land to the government to establish police stations, municipal offices, schools, and other government premises. The Swargashram Trust constituted the municipality in all respects, in terms of water supply and drainage, through the late 1970s. In 1976, the Jal Sansthan was established under the UP Water and Sewerage Act.[7] Since then, the Jal Sansthan has provided tap water to the facilities and homes in the Nagar. Fifteen years ago, the Trust began to shore up its assets and stopped ceding land to the government. As charitable trusts, all the large ashrams of Rishikesh enjoy a tax-exempt status and do not pay property, income, or goods and services taxes to the government. Ashrams are supposed to operate based on donations, meaning that visitors can pay whatever they can afford or want to, to stay there and use its services. Yet some of the large ashrams charge sliding fees with the best rooms reserved for the big donors.

WATER AND SEWERAGE COSTS

During our data-collection exercises, my colleagues and I found that we could not obtain information on water supplies and sewerage costs in every location. Some of the hotel and ashram owners or managers were not able to explain with precision their water and sewerage costs over time. There were also gaps in their knowledge of groundwater usage because their borewells were not metered and sewerage costs levied by the municipality for dumping septic wastewater into the city drains were variable across businesses. Hotel and ashram managers were also surprisingly vague in their estimates of the water consumption of visitors and pilgrims. In a few ashrams, the respondents did not provide information because it appeared that there was no sufficient recording, especially of water supplies and wastewater flows. Ashrams rely upon borewells for almost all their water needs, yet there are no flow meters. The largest hotels, by contrast, have water meters and must log daily wastewater production and consumption. Given these disparities, I used piecemeal data as I did for the five-star hotels in chapters 4 and 5, to trace what cost and consumption patterns might be. Table 2 represents the piecemeal data my research colleagues and I were able to collect and subjectively verify for water supplies and costs, sewerage services, STP installation, STP maintenance, energy costs, electricity costs, and other taxes.

At first, the perception of water scarcity appeared to be determined by the proximity to a groundwater source. In interviews and through the survey, owners and managers of hotels and ashrams who did not have a functioning borewell expressed water stress. While the hardship was less evident in Swargashram just along the river's edge, it was increasingly apparent among Tapovan's smaller hotels as our surveying moved away from the riverbank and above the main market road (Badrinath Road). Residents, guesthouse owners, and hotel managers residing farther up the hill felt that they suffered from water scarcity in the summer months. Those without a borewell had to purchase water from private tanker companies during the summer months.

The nagar of Swargashram and the village of Tapovan each have a sewage treatment plant on the edges of their neighborhoods, and these treat sewage from the centralized sewer lines of both areas. The sewage treatment plants (STPs) were installed by the state Pey Jal Nigam through state and central funds connected to the National Mission for Clean Ganga (NMCG) and urban renewal programs.[8] In Tapovan, the Pey Jal Nigam constructed the STP on land very near to the largest hotel in town. This hotel had been shut down by the courts and NGT many years earlier for dumping their wastewater off the back side of their property and into the River Ganga. The establishment of the new STP at Tapovan provided a place for the hotel to legally send their wastewater, and they were able to reopen before they built their own treatment plant.

Table 2 Costs Borne by Hotels and Ashrams for Water Supply, Water Procurement, Sewerage Services, STP Installation, and Maintenances, Energy, and Taxes

Ashram/Hotel	No. of Rooms	No. of Borewells	Water Supplies and Amounts	Energy Costs for Groundwater Pumping	Water Costs (price plus usage and procurement)	Sewerage Fees Charged by Government	STP Installation Costs with Pipe Costs	STP Maintenance Costs
Swargashram Trust	150	2	Borewell pumping 10–12 hours in summer and 4–5 hrs in rainy season	NA	NA	None	None	None
Parmarth Niketan	1,000	NA	NA	NA	NA	NA	None	NA
Geeta Ashram	800	5	Pumping 10–12 hours/day in summer and 4–5 hrs/day in rainy season	Rs. 36 lakhs annually and 7.5 hp pump costs	Almost none	Rs. 45 per toilet per month or Rs. 47,002 for 2 years	None	None
Vanprastha	200	1	NA	NA	NA	NA	None	None
Kailashanand Mission Trust	100	1	NA	NA	NA	NA	None	None

Saccha Dham	100	1	NA	NA	NA	NA	None	None
Hotel A	115	2–3	40–135 kld range; 69 kld average	Pump costs		Rs. 35 plus tax per toilet per month	4.5 lakhs	25,000 per month
Hotel B	22	2	Borewell pumping 3 hrs/day in peak		Rs. 700–1,000 for govt water per month	22 rooms x Rs. 35 plus tax per toilet/mo or Rs. 10–20,000 annually	None	None; septic
Hotel C	31	0	2 govt. connections totaling 50 kld		Rs. 4,800/mo per connection to Jal Sansthan	Rs. 35 plus tax per toilet per month	50 kld STP installed	NA
Hotel D	30	1	35–40 kld peak		Rs. 1,000/mo to Jal Sansthan	65k for connection	Old STP installed for 9 lakhs	None; offline Est 30,000/mo
Hotel E		2	80 kld peak–5 kld and recycled water for garden	Rs. 3,000–5,000/mo	Rs. 2,000/mo to Jal Sansthan		50 lakhs installed	25,000/mo; 16 lakhs annual

DECEPTIVE PRACTICES

During my discussions with hoteliers and residents, the mention of waste-water dumping in the Ganga raised apprehensions. The tension was just below the surface. One hotel manager alluded, "The authorities are very strict," as if to say that they can shut an establishment down at any time if they deem it to be noncompliant. To get an initial understanding of the number of hotels with STPs, I asked general questions. "How many hotels here have their own STP?" The questions were at first dull and muffled. Some were quiet. I continued, "Are they all connected to the sewer system, or do they have their own septic tanks? What is the current running load of the neighborhood STP at Tapovan?" First, businesspersons and residents answered in the negative: some said 30 percent are not connected to the STP; an engineer noted that 50 percent were not connected. As questions were directed specifically to hotel managers and ashram workers, it was apparent that hoteliers were aware of the orders and monitoring exercises involving STPs, but individuals working in management positions in ashrams were not. But both hoteliers and ashram managers were familiar with bans and rules related to water supply and were comfortable explaining the hardships in obtaining water.

In 2015, the NGT issued the key order that hotels and ashrams had to install sewage treatment plants.[9] The Pollution Control Board (PCB) then gave the notices to the three- to five-star hotels in Rishikesh, taking their authority, as they have in all other actions described in this book, from the Environment Act of 1986. However, the administration of these notifications to all the eligible units was uneven from the start. As an officer in a Nagar Panchayat in Rishikesh informed me, the PCB did not issue notices to any of the ashrams. Therefore, those ashrams with rooms numbering in the hundreds, and producing as much if not more wastewater than the hotels, were not given the notification to install STPs. Only one ashram, toward the end of my research, set up its own small treatment plant on their ghat as a "voluntary" action. When asked how they were motivated, they said from their philosophy and commitment to Ganga, and not because they were issued a notice by the NGT or the Central Pollution Control Board.

Turning now to the water and sewage arrangements in each of the large hotels and ashrams, short vignettes will convey their key features and variations. Following Table 2, Hotel A had 115 rooms with an average consumption of 750 liters per room per day for bathing and kitchen work. With their bore-well supply, they consumed an average of 69 kiloliters of water per day. They did not need to buy water from the Jal Sansthan. Hotel A said they ran their wastewater through an STP that they had just upgraded to treat 200 kiloliters a day. They claimed that they had spent 45 lakh (4,500,000) rupees to upgrade the system, but the bioreactor and the sand and carbon filters appeared unused. The hotel management contracted the installation to a Delhi-based firm but had decided to handle their own operation and maintenance with their own full-time staff. I met several of these staff members over several visits by just dropping in and saying hello. During those times, I could see that there was a full-time presence at the STP site.

The Hotel manager claimed that they directed 80 percent of their wastewater to the Tapovan STP nearby. But they also claimed to reuse 80 percent of their treated wastewater for gardening and cleaning. There was no evidence that they had run the plant since upgrading it in 2018, so the claims about treating and reusing their wastewater appeared specious. When explaining the costs, the STP manager explained that operating the STP for reuse would cost them Rs. 25,000 per month. When I visited in 2019, the STP staff member told me that the Pollution Control Board was checking their logbooks, electricity consumption, and flow meters, and the PCB was taking samples from the final storage tank to test on several parameters. But after pressing further, he admitted that the monitoring agents had not come in the last year to find out the results and assess the machine-microbe performance.

Moving on, the manager at Hotel B said, "At this point the NGT is very strong." This hotel had a septic tank but no STP and paid Rs. 10,000–12,000 to the Pey Jal Nigam in annual sewer charges. This payment permitted them to dump into the trunk sewer, which ran near their hotel and to the Tapovan STP a quarter of a kilometer down the road. The manager said, "I am star category, so I have to follow [the rules]. Small guest houses may have problems and not follow rules." Yet since they did not have an STP, it appeared they were not following the rule to install one. Instead, I guessed that their meaning of compliance referred to the practice of dumping their effluent from the septic tank into the public sewer line without pretreating it.

Hotel C did not have a borewell, but the management wanted one. Instead, they had two water connections from the Jal Sansthan (the municipal water board). They had installed dual plumbing in the entire hotel because the hotel owners had a pipe manufacturing business. When the manager showed me the dual system of hotel pipes they had installed on the back side of the outer wall of the hotel, they appeared unused. A year earlier, they had installed an STP with a capacity to treat fifty kiloliters per day and they claimed they had been running it for six months. From my visits on several occasions, I could see that the STP was operating and that the sewage was passing through the machine and then flowing into the main trunk line. It was not clear that treatment or digestion was occurring in the process since I was not shown samples from the final chamber. Usually when STPs are running well, the operators are interested in showing samples so that I can visually assess the water quality.

Hotel D claimed it built an STP before the NGT ruling in 2015. In 2018, the manager said that the NGT was inspecting the STP once every few months. On my one week stay in the hotel in 2019, I found that the STP was not operational. It was clear to me at that point that it had been offline for some time. Although the manager claimed that inspectors from the Pollution Control Board come every two to three months, there was no evidence of visits over that year. The manager deflected my focus: "See people don't pay taxes. Only 3 percent pay, and out of that we taxpayers are carrying it all. Poor people don't pay; ashrams make trusts to avoid taxes; people make a trust and then no taxes; ashrams are doing full business under the name of the trust."

RULEMAKING AND COMPLIANCE

These vignettes show that private businesses, public monitoring agencies, and the religious establishments are complicit in this system. They agree to build the infrastructure, but then feign compliance with rules and fabricate machine realities. In the following, I explain how this has been done in three phases. By dividing rule and policymaking, and responses and practices into three phases, I can show the shifts and changes taking place as legal orders reach the ground in Rishikesh. The following timeline also displays the power of the NGT to direct regulatory action for the government, even when noncompliance is hard for a visitor to see.[10]

Phase I, 2015–2016

In September 2015, the NGT issued the order that all hotels and ashrams with more than twenty rooms had to install their own STPs.[11] In October 2016, they issued a firm follow up. Before issuing these orders, the court required concerned government agencies to document all wastewater drains and effluent flows.[12] Along with these orders, the NGT directed that all government work related to sewer lines needed to be cleared with a court-appointed Principal Committee. It ordered: "If any work of this nature is to be carried on by the State of Uttarakhand or any of its Instrumentalities or Public Authorities or Bodies, it shall submit the proposal to the Principal Committee. The comments of the Principal Committee would be placed before the Tribunal for final orders. We also direct that no fresh works will be undertaken by the State or Public Authorities without approval of the Tribunal in relation to collection, treatment and disposal of the sewage except the works specifically provided in this judgement."[13]

These orders also followed with the direction that the STPs should eliminate fecal coliform bacteria and achieve in the treated effluent the parameters for BOD, TSS, and COD set by the CPCB. The NGT ordered the creation of a team of senior officers from the Uttarakhand Pey Jal Nigam (the state engineering group), the Uttarakhand Pollution Control Board, and the Department of Urban Development to submit quarterly reports to the Tribunal. The orders were issued by the NGT chairperson who had also issued the orders on drains in the Ganga River basin that I introduced in chapter 1.

After the NGT orders were issued, the Pollution Control Board gave "notices" (which are papers to enforce the orders) to the three- to five-star hotels to install STPs, but they did not issue the notices to the ashrams in Swargashram and Tapovan. All three- to five-star hotels with more than twenty rooms installed some kind of treatment machine to mark initial compliance.

Phase II, 2017–Mid 2018

By 2017, the NGT had grown more powerful in its monitoring of government bodies in the wastewater management field. In an important July 13,

2017, order, the NGT chairperson directed that an Executing Committee appointed under the judgment would be responsible for completing the upgradation of six government centralized STPs and would ensure that the projects were completed and operationalized within the time noticed in the judgment. The NGT reiterated its 2015 order:

> All the Hotels which have failed to establish their own STPs, and have failed to obtain the consent of UKPCB despite persuasion and public notice dated 15th September 2015 and are releasing their domestic waste and sewage into River Ganga or its tributaries and/or the drains whether or not leading to the STPs in Rishikesh or Haridwar, shall be directed to be shut down forthwith. The hotels which have applied for obtaining the consent of UKPCB in response to the above-mentioned public notice shall be granted and/or refused consent within 1 month from the date of pronouncement of this Judgement without default. Similarly, ashrams and dharamshalas which are discharging their sewage or domestic effluent directly into the River Ganga or its tributaries, whether or not they have their STP, would be directed to stop such discharge within 1 month from the date of issuance of the notice in this regard. Drains which directly bring sewage to the STP already established or to be established as afore-directed shall be connected to the common conveyor belt. The ashrams/dharamshalas which do not have their own STP would be required to establish such STP within 3 months from the date of pronouncement of this Judgement. They will not, in any event, be permitted to release their discharged sewage or domestic waste into River Ganga directly. They must discharge such effluent into drains alone that bring such effluent to the STP. If any hotel, dharamshala or ashram violates these directions it shall be liable to pay environmental compensation for causing pollution of River Ganga at the rate of Rs. 5,000 per day. The Joint Inspection Team referred above shall conduct inspection of the hotel, ashram and dharamshala and if any of them is found to be violating these directions and/or whose STPs/ETPs are either not functioning effectively or not releasing effluent within the prescribed limits then the inspection team shall submit the report to the Tribunal quarterly.[14]

There were many respondents listed in this case and they tried to evade or delay the enforcement of these penalties, fines, and closure notices. They could do so in the understanding that the chairperson was soon to retire, scheduled to step down in December 2017. After his retirement, the central government appointed an interim chairperson.

Phase III, Mid 2018–2019

In June 2018, a new NGT chairperson was appointed, and a series of leadership changes occurred. The new chairperson did not issue any follow-up orders for the Rishikesh hotels and ashrams. In response, most of the hotel managers I interviewed stopped running their plants if they had operated them earlier. At that point, compliance was watered down to mean having an STP but leaving it inoperative. I noticed this as our student research team visited the hotels in 2018. Many said to me, "Yes I have one" but it took me some time to realize that the statement meant that some kind of machine was there but not actually working. Nevertheless, the machinery allowed the management to declare compliance. Many of the hotel managers I interviewed intimated that this micro-action of fabricating machine reality was deliberate but temporally flexible. Their thought was that if the NGT came back again with more rigorous monitoring they could ramp up their machines to a partially functional state.

By late 2018, as the NGT's monitoring of Rishikesh was growing weaker, the Uttarakhand Pollution Control Board was easing off on its visits to hotels. When we conducted the survey in September 2018, I could see that the hotel managers knew they could slack off on the water-testing protocols for STPs. But the hoteliers were in a transition frame of mind, still fearful of NGT monitoring but stalling in bringing their STPs to a functional state. In visits to all the STPs of the large hotels, I found that two of the seven were operating their treatment plants. One hotel had an elaborate setup for using the treated water for gardening. At the same time, the central government was constructing three more centralized STPs as part of a regional infrastructure push that included a new railway line and the Char Dham Highways project.

In July 2019, I visited all the projects again. At Hotel A, the engineer said that the pumps used to draw the sewage from the septic tanks to the STP were broken and under repair. I asked about how long the repairs would take. "It could be a few days or a few weeks," he replied. I continued, "When was the last time the CPCB visited to check the system?" He returned, "It has been over a year." So I pressed, "Well how can I know if this STP has ever been running? I saw it being updated last September and now almost a year later I am wondering if it has ever been running." The engineer just smiled, without saying anything.

By mid-2019, monitoring by the PCB had stopped. When we started our discussions with hotel managers in 2018, they said that the PCB officers were coming on monthly or surprise visits. But after some pressing, several hoteliers admitted that PCB officers had not come in for over a year. While monitoring of these establishments was waning, the government and media were giving increasing attention to three new centralized STPs being built under the government's new Jal Shakti campaign.[15]

Phase III did not constitute the end of these STP rules by creating such an unviable situation that all parties agreed to give up. Hotel owners did not take operation of the machinery seriously, but they knew that they could ramp up their plants at a moment's notice if inspectors started arriving again. However, a broken or dysfunctional machine cannot be fixed rapidly, so there was the potential for other pretensions to unfold during an inspection visit by a monitoring agency. As one hotel owner intimated, "The NGT is still there." By keeping the STPs offline, hoteliers could save about Rs. 20,000 to 30,000 per month in operational and electricity costs. Meanwhile, the ashrams remained off the radar.

COMPLIANCE AND COSTS

There are several reasons for widespread noncompliance among hotels and public institutions such as ashrams. First, noncompliance evolves during times of weakened supervision from the NGT and the pollution control boards. At these times, hotel managers had little incentive to use their machines, since the end goal was to simply push the treated effluent back into the centralized sewer lines. Only one hotel was reusing their treated water for their on-site gardens and with enough groundwater for many hotels, compliance could not be assured by the interest in reuse. Simple compliance could be presented by the fact of having a machine, especially since the NGT was far away in Delhi and the local monitors were not appearing on site. Like those cases described in chapter 3, compliance with treatment machine standards declines when reuse is not an interest or option. This business noncompliance also leaks into public consciousness, where we found that hypothetical community acceptability for reusing treated wastewater was quite low.[16]

Weak compliance is also shaped by the scale of costs. The hotel management practices of spending economic capital by establishing a machine but then producing only symbolic capital as a pretend machine ended up being costly for these hotels, even before they add operation and maintenance costs to the mix. These systems did not produce usable water to meet some of their supply needs. The government-run STPs were also a deterrent to self-sufficiency, as business and ashram leaders knew they could discharge their effluents in the central sewer lines, which would then be treated in the neighborhood STPs.

Compliance and cost structures are also determined by the size of communities and businesses. Kuttuva et al. found that at two thousand onsite facilities in Bangalore, STPs declined in functionality and feasibility as the community's size decreased.[17] The smallest communities were unable to finance their facilities in a sustainable way because their dues were not sufficient to bear the STP costs and the maintenance and optimizations. Kuttuva et al. explained, "A combination of cost and enforcement make large apartment complexes more compliant than the smaller ones."[18] Smaller housing societies had more difficulties raising and sustaining all the capital, operational, and maintenance payments and to have the bioreactor and filtration systems needed to produce reusable water. Kuttuva et al. recommended that the smaller housing societies should be exempted from the requirement to install an STP.

Many cases in this book have shown that peri-urban circumstances are varied, with some communities and businesses connected to centralized water and sewer grids while others not. Therefore, the size of the housing society is one determining factor in shaping wastewater management responsibilities, but community or business size does not solve the problem of water scarcity where it exists. In water-scarce locations, wastewater treatment and reuse may even be viable options for housing societies with less than fifty apartments or flats. The consideration of compliance, costs, and community size must situate machines and functionality within the variable water availabilities in each community or business to understand how the determination is made on affordability.

The cases in Rishikesh also bring up the exemptions that some large institutions enjoy. While providing services to large numbers of visitors, leaders of ashrams and their managing trusts enjoy a special status with the state in terms of taxation, land ownership, and stewardship of critical water resources. Exemptions from the wastewater rules are part of the unspoken political agreements that many maintain with political leaders and parties.[19] Enforcing compliance to STP rules and bolstering monitoring of ashrams and large hotels could improve wastewater treatment practices and generate more reuse options.

7 Conclusions

Decentralized wastewater treatment systems are not a perfect solution in a world of pernicious wastewater flows. Yet these systems help to reduce in small steps the wastewater influx, which, on local and planetary scales, is endangering our fresh- or blue-water supplies. After treatment, wastewater can be reused to provide new supplies of water where there is water stress. This book has taken experiments with on-site community wastewater treatment and reuse in a positive spirit, outlining the sincere attempts as well as the insincere deceptions, the structural inadequacies and the courage of residents, professors, business leaders, nongovernmental consultants, activists, and judicial leaders to persist in addressing this challenge. The case studies have shown that experimentations with wastewater treatment and reuse significantly temper the disgust factor and lead to new territories of possibility for modulated and on-site activities. In this experimentation, humans use machines and microbes to digest their wastewater. In their human-machine-microbe interactions, they dynamically optimize and calibrate wastewater treatment and reuse to water availabilities and costs; site resources; community scales, rules, and policies; and bacterial affinities.[1]

This book has taken a human-machine-microbe perspective to account for the critical role of bacteria in the functioning of wastewater treatment and reuse systems. A human-machine-microbe perspective for the wastewater sector advances STS theory and human-machine approaches by including knowledge from microbiology, hydrology, and human governance. Without promoting and adjusting bacterial communities, wastewater cannot be digested, and machines become ineffective and inefficient. Effectiveness and efficiency are directly dependent upon the activities of microbes in primary and secondary treatment phases. STP operators must promote and assess microbial levels regularly to ensure optimum functioning.

Maintenance and repairs, two areas critical to the circular economy, involve microbes. Accumulations of microbe-laden sludge must be removed from baffled reactors from time to time. Ventilation systems must be cleaned to ensure the safe exhalation of the methane produced by anaerobic bacteria in these reactors. As residents encounter residual bacteria and suspended solids in their toilets, they get accustomed to "brackish water" or request further optimization to improve microbial digestion of biomass in the water. Discomforts produced by smell or pollution evident in appearance motivate tertiary applications such as chlorine dosing or dilution with rainwater. Communities of microbes must also be killed off in tertiary treatment with chlorine or UV light so that waters can be safely reused. By accounting for the role of microbes, an STS theory can be attentive to the ways machines and infrastructures intersect with the biological world.

DECENTRALIZED AND NESTED LOOPS
IN THE HYDROSOCIAL CYCLE

In a human-machine-microbe perspective, the framework of the hydrosocial cycle tracks citizens' views of water transformations from unusable to usable states, which are then considered for a range of non-potable purposes. Water is not seen as a homogeneous stream; there are many kinds of waters, and their usabilities are distinguished according to technologies and human needs. In the category designated as non-potable waters, some treated waters are good for horticulture, while others can be used for toilet flushing. Others are good enough for AC cooling towers. The distinctions on water types and their purposes are dynamically calibrated as STP operators and citizen and business water users experiment with sanitation machines. Water distinctions help to forward the options for wastewater reuse.

The best examples of wastewater reuse show that decentralized projects are creating nested loops within larger hydrosocial cycles. I borrow the metaphor from computer science, where it designates iteration and repetition in software coding, and from the nonlinear nested cycles of social-ecological systems.[2] Decentralized hydrosocial loops of wastewater treatment and reuse are nested in wider centralized systems of water and wastewater circulation, even if they are not perceived that way by users. In cluster housing, five-star hotels and large colleges and universities, the nested systems draw potable and non-potable water from various broad cycle systems such as groundwater from different locales transported by tankers, rainwater, and piped water from rivers and reservoirs. The uses then fulfill human needs at community scales and the smaller nested loops attempt to keep as much of this water as possible on-site. The externalization of wastewater to other communities is reduced through on-site reuse. The greater and more prolific wastewater flows existing outside of these looped locales can also be tapped to produce usable water on-site. In garden STPs, wastewater was extracted from the centralized hydrosocial cycle and brought a short distance to create a smaller loop of reuse water. In this cross-scale nesting of loops, the centralized sewer system can be utilized to build smaller hydrosocial loops of wastewater reuse within the context of a garden, a business, a university, or a housing society. Garden STPs put treated wastewater back into the soil where the soil's oxygen and carbon digest the wastewater further. This kind of loop augments the soil moisture of gardens and reduces the need for groundwater extraction.

The understanding of nested hydrosocial loops and cycles brings the sewershed into focus. Sewersheds are made up of toilets, sewer pipelines, drains, and treatment plants. Housing complexes, five-star hotels, and colleges and universities use their own sewersheds as the scales for experimentation and project activity. The sewershed adds to understandings of decentralized infrastructures that exist in the gaps between formal, centralized grids, at the edges of centralized grids, or in spaces within centralized grids where a government or business group determines that extra attention to flows is needed. In their varied placements, decentralized projects can tap into the flows of wastewater in centralized sewers and create smaller treatment systems without being obstructed by the existing infrastructures. The garden STPs draw their wastewater sources from the centralized sewer grids of the city where the flows are hefty, and this reduces the load on centralized STPs. However, housing society members and small hoteliers may also maintain their dependence upon centralized sewer systems and avoid running their decentralized projects, which limits reuse potential. When peri-urban housing communities obtain new connections to centralized freshwater supply pipelines, they may then reduce their groundwater extraction and shelve their reuse aspirations. The greatest levels of commitment for continued reuse activities exist in the off-grid middle- to high-income communities, where connections to centralized water or sewer lines do not exist and in businesses such as five-star hotels that have high freshwater bills from the state. These off-grid communities and water guzzling businesses can visualize the possibilities and actualities of closed loop systems of reuse, even if they are not able to fully achieve them.

RECYCLED WATER ON THE SCALE OF WATER COSTS

The affordability of water is a rising concern around the world and especially among low-income and marginalized groups.[3] Within decentralized systems, costs vary by technology and the extent of water reuse. Simple systems with primary up to secondary treatment using a bioreactor and a planted filter or vortex can suit horticulture reuses. With additional filtration devices, water can be used for toilet flushing, street cleaning, and other outdoor functions. With even more advanced bioreactors with aeration and media devices to aid microbial digestion, water can be produced for AC cooling towers. Each treatment method involves the costs for bioreactors, filtration devices, pipelines, and ancillary equipment such as pumps. Companies and communities generally choose the technology and treatment level based on their budgets and water costs from other supplies.

New producer and consumer roles, simultaneously producing and consuming treated wastewater within closed or small loops, are displacing or sidestepping the state pricing mechanisms and creating new markets for water. RWAs are using a portion of their monthly maintenance fees, which range from Rs. 2,000 to 6,000 per household, to build or upgrade their STPs, maintain them, hire managers, or contract with a maintenance company. Businesses, universities, and colleges are paying for their systems through their operational funds. Five-star hotels are able to recoup some of their costs through savings in their water bills. Some office buildings and universities had estimated that their on-site projects could deliver treated water for reuse within a better cost structure than the higher-priced rates for commercial buyers and tanker and ATM water. Those that did not save much on their water bills saw benefits in other ways. They valued the backup water supply, or they felt good about saving freshwater for potable and household uses. Or they foresaw a future benefit in their water budgets after the initial installation costs were paid off. Given the reality of having less water, many were willing to spend more to generate new sources if their machines were up to the task. Consumers living off water and sewer grids in peri-urban areas and paying higher prices for water showed more interest in and motivation to reuse treated wastewater.

In government institutions, affordability in terms of cost per kiloliter is assessed regularly. In 2013, Evans calculated that it appeared beneficial for the BWSSB to provide customers with treated wastewater rather than importing freshwater from increasingly distant sources such as the Cauvery River. He wrote, "The cost to BWSSB of supplying treated river water is approximately 26–46 INR [Rupees] per kiloliter (BWSSB unpublished figures) compared to 10–12 INR/kiloliter for the treatment of wastewater."[4] Evans estimated that the revenue derived by BWSSB from the sale of treated water could reach around Rs. 5,000,000 per year. Evans's estimate of the cost per kiloliter to produce reuse water appears too low, but he is basing that upon centralized, large-scale STPs that may produce a greater volume of water.

For housing societies, businesses, and institutions in India, estimations of the cost per kiloliter for reuse water have not yet been made in any systematic way. Compiling the benchmarks provided in this book, a few initial pictures on relative costs can be made for these groups. Figure 22 shows the costs of water sources including recycled water, as respondents mentioned them in this ethnography. This visualization of benchmark costs can help to promote use of recycled water with the understanding that an

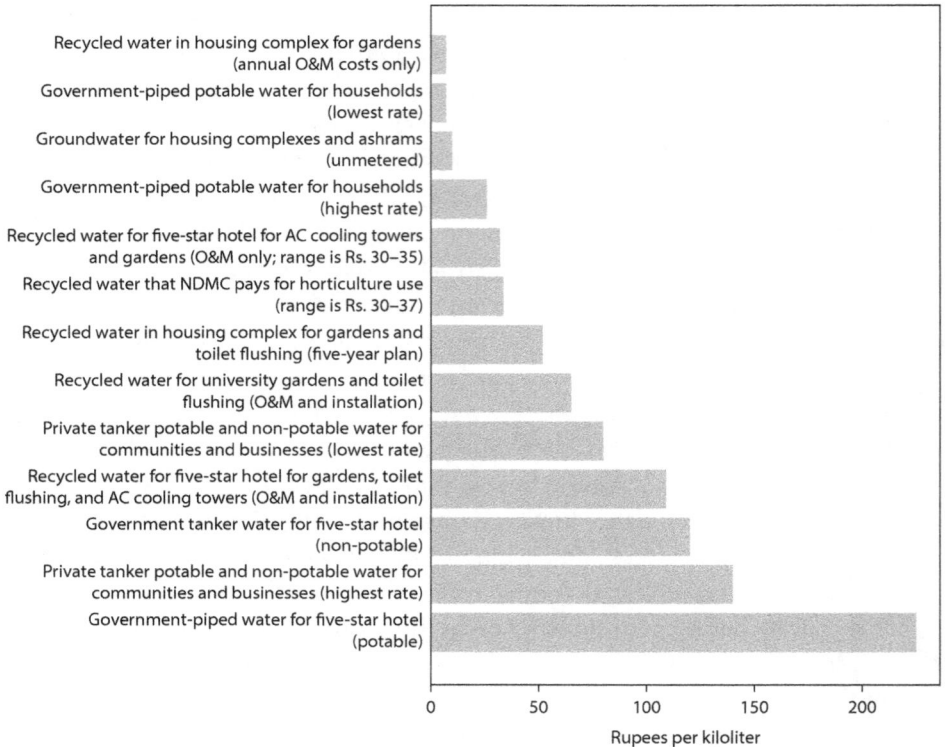

Recycled water in housing complex for gardens (annual O&M costs only)
Government-piped potable water for households (lowest rate)
Groundwater for housing complexes and ashrams (unmetered)
Government-piped potable water for households (highest rate)
Recycled water for five-star hotel for AC cooling towers and gardens (O&M only; range is Rs. 30–35)
Recycled water that NDMC pays for horticulture use (range is Rs. 30–37)
Recycled water in housing complex for gardens and toilet flushing (five-year plan)
Recycled water for university gardens and toilet flushing (O&M and installation)
Private tanker potable and non-potable water for communities and businesses (lowest rate)
Recycled water for five-star hotel for gardens, toilet flushing, and AC cooling towers (O&M and installation)
Government tanker water for five-star hotel (non-potable)
Private tanker potable and non-potable water for communities and businesses (highest rate)
Government-piped water for five-star hotel (potable)

0   50   100   150   200

Rupees per kiloliter

Figure 22. Costs of various water sources, including treated wastewater, on the basis of rupees per kiloliter. (Created by Ali Krzton)

estimated cost per kiloliter must be situated in the larger frame of all water costs incurred by a household, business, or institution. New financing arrangements appear in deals between government agencies, companies, consultants, and community associations. For the garden STPs and the STP for the bus depot in Delhi, capital and operational costs were divided across government departments and private companies. In the IIT-M case, university operational and research funds enabled the establishment of the campus STP and the monitoring tools for research. In the Marriott case, the hotel invested in upgrading and expanding their infrastructure using private companies. This promotes innovation in decentralized treatment systems but puts new pressures on businesses, universities, RWAs, and other community groups to organize their own wastewater services. Kundu has shown that this transfer of responsibility from the state to the RWA is not new.[5] RWAs have been centrally involved with water provisioning for a longer period of time. But the added responsibility of provisioning the wastewater public good, in addition to providing for the water public good, is stretching community and business capabilities. Low-income communities without strong local collective action suffer from the double burden of water and wastewater provisioning and end up giving priority to water supply as sanitation investments languish.

SELF-MONITORS AND THE CHALLENGES OF THE CIRCULAR ECONOMY

This ethnography has given visibility to new ways of provisioning the public goods of water supply and wastewater treatment that have implications for governance and municipal functions and public-private partnerships. Regulations created by the courts, the NGT, water and sewer boards, and the pollution control boards have shifted decision-making on technologies from government agencies to communities and businesses. New roles in technology uptake are created among apartment builders and Resident Welfare Associations (RWAs). Residents within their RWA boards make decisions on technologies by soliciting and using input from NGOs and science advisors to tweak the infrastructure provided by the builder or choose another design that builds on resident interests. They choose from a variety of bioreactors and filtration devices. Alterations in design are made by these groups to optimize digestion and prevent odor and gas. Hotel owners and managers upgrade their STPs to perform higher-functioning tasks that produce bulk water for essential services such as air-conditioning. University researchers build machines in campus labs in search of the optimum mechanics for achieving a BOD of 10. Knowledge of technologies and awareness about treatment phases and outcomes traverses horizontally across RWAs, hotels, universities, and their NGO consultants in a rhizomatic fashion, creating new methods of governance alongside the government's unwieldy water bureaucracy.

While municipalization has characterized the last half of the twentieth century in India, the twenty-first century has seen a shift toward privatization in petroleum, hydropower, power renewables, airlines, railways, and shipping.[6] Critics have pointed to the influence of neoliberal agendas promoted by developed countries and their international aid agencies in this shift to greater private-sector involvement. The rise of the middle class in India since the 1990s has also generated human resources for the development of private companies and nongovernmental agencies. The burgeoning middle class has spurred changes in urban governance with the proliferation of RWAs and other religious and secular associations."[7] Harriss has noted that liberalization requires the "shrinking of the space of state action and the devolving of functions to the private sector or to civil society, whilst still needing instruments of rule." Civil society participation and community involvement are instrumental in reconciling the tension between the different objectives of state and society because they presuppose or are expected to encourage the development of "self-rule" and of people's capacities to look after themselves and their communities. He explained further, "They presuppose the consumer-citizen subject who looks for accountability and efficiency, but who also submits to the disciplines that they require."[8] Using this lens on the post-liberalization state, self-monitoring of water quality to meet community reuse needs looks like a deepening of neoliberal agendas on property ownership, cost structures, and regulatory kickbacks, and marks the abrogation of the state. But these measures are also stimulating the circular economy of water where resident assessments are accounting for resource availabilities, stresses, and common incentives for conservation and the reduction of waste.

RWAs, businesses, builders and nongovernmental organizations are capturing benefits from circular water reuses within the neoliberal functioning of state and nonstate agencies, by self-monitoring technologies and water qualities using tighter feedback loops. Tight feedback loops, representing the paths by which decisions, behaviors, and energies flow, transmit assessments of accountability. Accountability for proper STP functioning was assessed internally according to experiences with reuse and not externalized to government agencies who were absent from assessing reuse water qualities. In tight feedback, there is a short path between the operational decisions and behaviors within a facility and the user's response to the system. In some cases, businesses are motivated to self-monitor by the need to achieve the higher water-quality levels for toilet flushing and AC cooling towers. Water-quality monitoring is driven by citizen and business experiences, which then feed back to the STP managers for their use in optimization. The tighter the feedback loop, the stronger the acceptability for reuse will be. Tight feedback also means that input into the system may be limited to the actors in the loop.

The directions and alliances of privatization and remunicipalization are mixed, as remunicipalization includes the emergence of community-based circular economies. Each project is created out of a participatory tangle of consultants, government agencies, private companies, and resident associations. The courts and pollution control boards direct company and community controls over water in the decentralized wastewater sector by slapping on mandates for STPs. Users must establish their machines and engage microbes for any public good to come of it. Opportunities to develop reuse water can reduce other water costs. But opportunities are unevenly captured and distributed, as middle-class participation caters to middle-class communities and does not ensure the needs or participation of the urban poor. This is a dilemma raised by those working to advance the circular water economy.[9] However, there is scope for increasing participation, as monthly fees are raised by resident communities and businesses representing a broader spectrum of income levels. In these experiments emerge the potential to generate decentralized projects that suit community needs.

The deepest governance challenges for wastewater management and reuse arise from deceptive practices stemming from the cunning state. But in cases where deceptive machines have prevailed, the state has not been the only cunning actor. Other agencies and actors have been involved in cunning practices. Government agencies will shirk their responsibilities by creating the pretense of functioning STPs that are in fact dysfunctional or nonfunctional. In housing projects, these pretenses keep residents stuck in the muck, while in small hotels the pretenses save long-term costs. But among five-star hotels, high water prices mitigate against noncompliance by incentivizing the production of new water supplies from reuse water. While more time is needed to assess the effectiveness of universities, colleges, hotels, and RWAs at delivering these public goods of reuse water, communities of users are driving new modes of accountability as they evaluate the effects and benefits of the funds they spend on wastewater treatment.

RECOMMENDATIONS

One of the key insights that can be drawn from looking across cases in low-, middle- and upper-income communities is that those who can raise more monthly funds through their RWAs are better off in terms of capacity to produce reuse water. This leaves the looming question of how lower-income communities can be brought into the fold or helped by decentralized wastewater systems and reuse options. If low-income communities are not able to raise their own monthly funds for sanitation, then they are beholden to governmental and nongovernmental interests and decisions that work ideologically to promote DEWATS systems. The challenge is to find ways to make such systems work sustainably.

In thinking about low-income communities, I can offer several recommendations. First, low interest micro-loans could be provided to communities that have treatment plants but want to upgrade them to produce reuse water. Examples of neighborhood businesses producing RO, or reverse osmosis, water have been referenced in one of the examples in chapter 3. It may be possible to extend that entrepreneurial spirit to develop RO systems using treated wastewater as the input. Given the cost of tanker water that many low-income communities rely on, there may be scope for interest in generating systems that can produce water for toilet flushing. There is also a great need for demonstration projects that produce water or other benefits in low-income communities. The CDD demonstration STP in another case in chapter 3 did not produce visual reuse results, so residents could not imagine their involvement and the benefits in terms of producing another water supply. Workforce training in this sector can be aligned with community needs and youth interests in employment. Other kinds of innovative financing strategies to support investment in water infrastructure projects have been suggested in the United States and could be applied in India.[10]

Other recommendations related to sustainability, community acceptance, participation, and technology experimentation can be made for all communities. On-site water and sanitation planning are necessary for the success of these projects and RWAs could benefit from guidance manuals to strengthen their decision-making. Since RWAs already pay for regular cleaning of drains, tanks, pipelines, storage tanks, and bioreactors, financial incentives and tax breaks could assist in the sustainability of these tasks. Tax incentives can be provided by local and state governments to ensure that bioreactors and filtration devices can be purchased, upgraded, and staffed. Government buy-back of community-produced reuse water could also be considered, on the model used for buying back solar power. A tax rebate on STP installation can also ensure the first-phase compliance of the rules mandating STPs and promote ongoing experimentation with technologies.

Community training for wastewater infrastructure can go a long way toward facilitating the democratization of water and sewer services. RWAs and their staff would benefit from on-site training opportunities, not just with established NGOs but with municipal bodies. On-site recertification training can be replicated on a wider scale. Training can also strengthen community acceptance and participation. Community experimentation and tight feedback loops generate the greatest degree of acceptance and willingness to use treated water. Exchanges of experience among housing community members can support awareness and decision-making and generate new ideas. Information on water monitoring and machine repairs can also be expanded through connections among RWA groups digitally and in person. Members can build and display their databases on best practices. RWAs and businesses can also disseminate their experiences with public-private partnerships to provide models of best practices. PPP models that build capacity for community members to earn income by selling treated water back to the municipality or to create fishponds and other food supplies could produce on-site benefits.

Connections with centralized infrastructures can be advanced if government agencies are willing to share infrastructure maps in digital form so that community groups can visualize their sewersheds. If viewed as resources, wastewater can be drawn from these centralized drains and used in community-based hydrosocial loops for the on-site production of nonpotable water. However, as Wright et al. note, there may be public health risks with this broadening of the circular economy.[11] There is a gap in understanding of the public health risks involved with drawing wastewater from large, centralized drains and treating it in community or business settings where people live and work. Antibiotic-resistant bacteria can spread if not disinfected properly. COVID-19 RNA may survive in untreated wastewater, even if it is not transmissible from the fecal-to-oral route, and its presence may create other problems not yet understood.[12] Monitoring devices needed to check for contaminants, and viruses are more expensive and may not be affordable for most communities and small businesses. More research is needed to address these public health concerns.

Additionally, improvements to decentralized management may be limited by the ongoing cunning practices of state and nonstate actors, which could undermine the effectiveness of the NGT and the pollution control boards as lead monitors. Deceptive or pretend machines can also continue the wastage of technological and economic capital and stunt the benefits of adding reuse water to the list of affordable water supplies. Companies can also renege on their contractual obligations in partnerships with communities and government agencies, wasting capital investments and eroding trust in collective action. Building community vigilance mechanisms targeted at pretenses and deceptions can go a long way toward preventing waste and empty promises.

Wastewater flux is a part of local and global water cycles. Capturing influxes and transforming them into usable waters through on-site systems can alleviate water stress in India and other parts of the world. If these kinds of projects are multiplied around the world, the use of treated wastewater will save freshwater and have a noticeable effect on groundwater tables and green water. Inspired by community-level action, solutions can generate more effective technological methods over time. Knowledge of the complexities of human-machine-microbe interactivities will support the creation of new kinds of usable water close to home and work.

Glossary

AC cooling towers   Heat exchangers that use water and air to transfer heat from air-conditioning systems to the outdoor environment. They remove heat from the condenser water leaving a chiller. Cooling towers are usually located on rooftops or other outdoor sites.

aeration   Adding air to a treatment process.

aerobic bacteria   Bacteria that thrive and grow in an aerobic environment (meaning "with oxygen").

Agra Nagar Nigam (ANN)   The municipal agency for the city of Agra in the state of Uttar Pradesh.

AMRUT (The Atal Mission for Rejuvenation and Urban Transformation)   An agency of the Central Government of India.

anaerobic bacteria   Bacteria that do not live or grow when oxygen is present.

anaerobic baffled reactor (ABR)   Developed in the early 1980s, a reactor that consists of a container with a series of compartments (up to eight) that are baffled to force the incoming wastewater through a series of sludge blankets that remove and digest organic matter.

antibiotic-resistant bacteria   A subset of antimicrobial resistance (AMR), it refers to bacteria that become resistant to antibiotics. Infections due to AMR cause millions of deaths each year.

Article 32   The Constitution of India, 1949, ensures rights by this Article that include: "(1) The right to move the Supreme Court by appropriate proceedings for the enforcement of the rights conferred in the Constitution. (2) The

Supreme Court shall have power to issue directions or orders or writs, including writs in the nature of habeas corpus, mandamus, prohibition, quo warranto and certiorari, whichever may be appropriate, for the enforcement of any of the rights conferred. (3) Without prejudice to the powers conferred on the Supreme Court by clause (1) and (2), Parliament may by law empower any other court to exercise within the local limits of its jurisdiction all or any of the powers exercisable by the Supreme Court under clause (2)."

ashram   Place of religious learning and practice for Hindus.

ATM water   Purified water purchased from a roadside unit.

BBMP (Bruhat Bengaluru Mahanagara Palike)   The municipal agency for the city of Bengaluru or Bangalore.

BDA (Bengaluru Development Authority)   A city agency that handles infrastructure for the city of Bengaluru.

bioreactor   Vessels or tanks in which whole cells or cell-free enzymes transform raw materials into biochemical products and/or less undesirable by-products.

BOD (biological oxygen demand)   The amount of oxygen consumed by bacteria and other microorganisms while they decompose organic matter under aerobic conditions at a specified temperature.

BORDA (Bremen Overseas Research and Development Association)   A network of wastewater professionals in Germany.

borewell water   Water from a deep well.

BOT (build-operate-transfer)   A process in building and operating public-private infrastructures.

BWSSB (Bengaluru Water Supply and Sewerage Board)   A city agency that handles water and sanitation for the city of Bengaluru.

capex   Capital expenditures for a project.

carbon filtration   A method of filtering that uses a bed of activated carbon to remove impurities from a fluid using adsorption.

CDD (Consortium for DEWATS Dissemination)   A network of agencies working on decentralized wastewater treatment systems.

Central Ground Water Authority   The main central or federal agency governing groundwater in India.

Chennai Metro Water and Sewage Board (CMWSB)   The water and sewerage board for the city of Chennai or Madras.

clarifier   Settling tanks built with mechanical means for continuous removal of solids being deposited by sedimentation. A clarifier is generally used to remove solid particulates or suspended solids from liquid for clarification and/or thickening.

closed loop   The recycling of gray and black waters repeatedly through the user system without being discharged back into the environment.

COD (chemical oxygen demand)   A test that measures the amount of oxygen required to chemically oxidize the organic material and inorganic nutrients, such as ammonia or nitrate, present in water.

common effluent treatment system   A wastewater treatment system owned and operated by a group of entities, such as industries.

crore   10,000,000 rupees.

CURE (Centre for Urban and Regional Excellence)   A not-for-profit development organization working with urban informal and low-income communities to un-think, reimagine, innovate, and de-engineer solutions to include and integrate people in the processes of city development.

DDA (Delhi Development Authority)   A planning authority created in 1957 under the provisions of the Delhi Development Act "to promote and secure the development of Delhi."

Delhi Jal Board   The water board handling water supply and wastewater management for the national capital territory of Delhi.

Delhi Pollution Control Committee (DPCC)   A regulatory body for the National Capital Territory of Delhi for the implementation of various environmental/pollution control laws enacted by the Parliament and notified by the Ministry of Environment, Forests, and Climate Change, in the Government of India.

desludging   Removing sewage from a septic tank or bioreactor.

dharamshala   Place of religious learning and teaching.

digestion   Process of metabolizing biomass by bacteria.

Directive Principles of State Policy   Constitutional principles that aim to create social and economic conditions under which citizens can lead a good life. They also aim to establish social and economic democracy through a welfare state. The Directive Principles of State Policy are guidelines for governance that the state is expected to follow in framing policies and passing laws.

DPCC (Delhi Pollution Control Committee)   The body that regulates pollution for the capital territory of Delhi.

dual plumbing   Two sets of pipes in a building, one for purified drinking water, and one for treated wastewater.

fecal sludge management   Management of sewage from septic tanks and pit and bucket latrines.

Forty-Second Amendment   An important amendment to the Indian Constitution that attempted to reduce the power of the Supreme Court and the High Courts. It laid down Fundamental Duties for citizens and added the terms socialist, secular, and integrity to the Preamble.

Fundamental Duties   Duties that define the moral obligations of all citizens.

fundamental rights   Seven rights originally provided by the Constitution: the right to equality, right to freedom, right against exploitation, right to freedom of religion, cultural and educational rights, right to property, and right to constitutional remedies.

Ganga Action Plan   The Government of India's first environmental program to reduce and prevent the pollution of the River Ganga.

Gram Panchayat   The decision-making unit for a village.

gravel or plant filters   A gravel bed through which wastewater flows and acts as a filter medium. Plants may grow in the gravel, sustained by the nutrients in the wastewater.

High Court   The highest court in each state of India.

"honey-sucker"   An individual or company that removes solidified sewage from a septic tank.

Jal Sansthan   State water board handling drinking water treatment and purification.

kld   Kiloliters per day, a measure used to denote the volume of wastewater that can be treated in a treatment plant, where 1 kld equals 1 cubic meter per day.

lakh   100,000 rupees.

mandamus   Legal remedy to hold an agency accountable for the work it is supposed to do.

MBBR (moving bed bioreactor)   A process used for the removal of organic substances, nitrification, and denitrification. The MBBR system consists of an activated sludge aeration system where the sludge is collected on recycled plastic carriers. These carriers have an internal large surface for optimal contact for water, air, and bacteria. MBBR is a biological treatment process based on a combination of conventional activated sludge process and biofilm media.

MBR (membrane bioreactor)   A combination of membrane processes like microfiltration or ultrafiltration with the activated sludge process. The MBR technology requires very little land but involves aeration to stimulate bacterial multiplication and thus uses more power. The unit contains an initial chamber where grit and solids are removed and a tall metal tank where the water is then aerated. In this open-top tank, the wastewater is infused continuously with oxygen to spur the multiplication of aerobic bacteria that then digest the biomass in the water. The clearer water runs off from that tank and through a membrane filter. This membrane filter removes remaining bacteria and particles and produces water usable for horticulture. Sludge is generated from this process and must be removed from the tanks and distributed for horticulture use.

membrane filtration (MF)   A pressure-driven separation process that employs a membrane for both mechanical and chemical sieving of particles and macromolecules.

methane   Gas produced by anaerobic bacteria.

Ministry of Drinking Water and Sanitation   A ministry of the Government of India formed in 2011. After May 2019, the ministry was merged with the Ministry of Jal Shakti.

Ministry of Environment, Forests and Climate Change   The ministry of the Central Government of India that covers all environmental matters as well as forests and climate change.

Ministry of Jal Shakti   Formerly the Ministry of Water Resources, River Development and Ganga Rejuvenation, this ministry oversees all water development and infrastructures in the country.

mld   Million liters per day, a measure used to denote the volume of wastewater that can be treated in a treatment plant, where 1 mld equals 1,000 cubic meters per day.

Nagar Panchayat   The decision-making unit for a town.

National Capital Region (NCR)   A planning region centered around the National Capital Territory (NCT) of Delhi. It encompasses Delhi and several districts surrounding it from the states of Haryana, Uttar Pradesh, and Rajasthan.

National Mission for Clean Ganga   The agency governing the Ganga Action Plan and all projects including wastewater treatment in the Ganga River basin.

NDMC (New Delhi Municipal Corporation)   The municipal agency for New Delhi.

NEERI or CSIR-NEERI (National Environmental Engineering Research Institute)   A research institute created and funded by the Government of India.

NGT (National Green Tribunal)   India's first environmental tribunal for hearing environmental cases.

Niti Aayog   The Planning Commission for the Government of India.

O&M costs   Operation and maintenance costs of a wastewater treatment plant.

opex   Operational costs of a wastewater treatment plant.

pathogen   A bacterium, virus, or other microorganism that can cause disease.

phytorid bioreactor   A constructed wetland exclusively designed for the treatment of municipal, urban, agricultural, and industrial wastewater based on specific plants, such as elephant grass (Pennisetumpurpurem), cattails (Typha sp.), reeds (Phragmites), Canna and yellow flag iris (Iris pseudacorus), normally found in natural wetlands with filtration and treatment capability. Some ornamental as well as flowering plants species such as golden duranta, bamboo, Nerium, and Colocasia can also be used for treatment as well as for landscaping purposes. The phytorid technology can be constructed in series and parallel modules or cells depending on the land availability and quantity of wastewater to be treated. The cell/system is filled with porous media such as crushed bricks, gravel, and stones. The hydraulics are maintained so that wastewater does not rise to the surface.

PIL (public interest litigation)   Litigation in which a citizen can stand for the public and bring a case forward in court in the public interest.

pit latrine   A type of toilet that collects human feces in a hole in the ground.

planted gravel filter or planted filter   The common reed used in a planted sewage filter bed develops micro-organisms that digest the pollutants in sewage.

This bacteria's development is achieved by transferring oxygen from the plant leaves down through its stem and roots resulting in bacterial growth in the gravel bed. The sewage effluent flows through this gravel bed and the micro-organisms treat the pollutants. Reed bed systems can be 100 percent non-electric if the site has a reasonable gradient; however, the use of electric pumps may be needed to lift the water in "flat" areas.

Pollution Control Board (PCB, central and state)   The main agency for environmental regulation and for the control and measurement of pollutants in the environment.

primary treatment   A treatment process designed to remove organic and inorganic settleable solids from wastewater by the physical process of sedimentation.

public-private partnership (PPP)   A collaborative effort for establishing an infrastructure or some other kind of project.

raja kaluve   An open drain.

remunicipalization   The process of returning the management of water and sanitation to municipal authorities after management by a private company or corporation.

retention time   The amount of time needed in the bioreactor to allow bacteria to digest the biomass in the wastewater influent.

reverse osmosis   A technology that removes contaminants from water by pushing the water under pressure through a semi-permeable membrane.

RGRHCL (Rajiv Gandhi Rural Housing Corporation Ltd.)   A government organization working in the field of housing.

root zone treatment chamber   A chamber in which root zone treatment purifies wastewater as it passes through an artificially constructed wetland area. The pollutants are removed by various physical, chemical, and biogeochemical processes like sedimentation, absorption, and nitrification through uptake by wetland plants. The chamber is similar to a planted filter.

RWA (resident welfare association)   An association that represents members of a housing community such as an apartment or condo complex.

sand filtration   A process in which wastewater is run through a sand filter to remove suspended matter, as well as floating and sinkable particles. The wastewater flows vertically through a fine bed of sand and/or gravel. Particles are removed by way of absorption or physical encapsulation. If there is excessive pressure loss on the filter, it must be rinsed.

secondary treatment   The second step in most waste treatment systems during which bacteria consume the organic parts of the waste. This is accomplished by bringing the sewage, bacteria, and oxygen together in trickling filters or within an activated sludge process.

septic tank   A tank, typically underground, in which sewage is collected and allowed to decompose through bacterial activity before draining by means of a leaching field.

sequential batch reactor (SBR)   An activated sludge process designed to operate in a batch mode with aeration and sludge settlement occurring in the same tank.

settling tank   A sedimentation tank that allows suspended particles to settle out of water or wastewater as it flows slowly through the tank, thereby providing some degree of purification.

sewershed   An area over which wastewater flows that includes drains, pipes, treatment plants, and surrounding waterbodies.

sludge   Semiliquid waste that is obtained from the processing of municipal sewage, often used as a fertilizer or as bacterial liquor for treatment.

Smart Cities Program or National Smart Cities Mission   An urban renewal and retrofitting program by the Government of India with the mission to develop smart cities across the country, making them citizen friendly and sustainable.

soil biotechnology method (SBT)   A low-energy wastewater treatment process that uses a trickling filter and layers of gravel, rocks, or other materials to contain the bacteria.

suo motu powers   Actions taken by a court of its own accord, without any request by the parties involved.

Swachh Bharat Abhiyan (Clean India Mission)   The Government of India initiative to improve sanitation across the country through a number of projects and programs.

tanker truck   A large truck that carries purified water or treated wastewater to locations of water need, or a large truck that carries sewage that has been removed from septic tanks.

TDS (total dissolved solids)   A measure of the dissolved combined content of all inorganic and organic substances present in a liquid in molecular, ionized, or micro-granular (colloidal sol) suspended form. TDS concentrations are often reported in parts per million (ppm). Water TDS concentrations can be determined using a digital meter.

tertiary treatment   The third and final advanced treatment process used to disinfect water that has already been treated by primary and secondary processes for removing harmful material in a wastewater plant. The process removes pollutants not adequately removed by secondary treatment, particularly nitrogen and phosphorus, using sand filters, microstraining, or UV light.

TSS (total suspended solids)   Waterborne particles that exceed 2 microns in size.

UKPCB   The Pollution Control Board for the state of Uttarakhand.

UK Pey Jal Nigam   The Sewerage Engineering Board for the state of Uttarakhand.

UP Jal Nigam   The Sewerage Engineering Board for the state of Uttar Pradesh.

upflow anaerobic sludge blanket (UASB)   An anaerobic process in which a blanket of granular sludge is formed that suspends in the tank. Wastewater flows upward through the blanket and is degraded by the anaerobic microorganisms. The upward flow combined with the settling action of gravity suspends the blanket with the aid of flocculants.

vortex   An aeration device that provides oxygen saturation in wastewater. It mimics a controlled tornado effect, and oxidates through a looping cycle. It also eliminates smell.

zero liquid discharge or ZLD   The notion that wastewater can be fully reused and not discharged into the environment even after treatment.

Notes

INTRODUCTION

1. Alley, Maurya, and Das 2018.
2. In 2016, only one company had produced a viable decentralized sewage treatment facility in Delhi.
3. Al-Saidi 2021.
4. Fielding, Dolnicar, and Schultz 2018.
5. Scruggs and Heyne 2021.
6. Barnes 2014; Walsh 2018.
7. Pasmore et al. 1982 and 2019; Ruiz-Quintanilla et al. 1996. For recent research on sanitation using the sociotechnical approach, see Sutherland et al. (2021).
8. Haraway 1984. See also Wells 2014.
9. Rose 2007.
10. Braun 2007; Cavanagh 2014. Kochhar's (2020) analysis of Hindu imaginations of the River Ganga's antibacterial properties is a unique case combining faith, filth, and the bacteriophage. He explores how the bacteriophage virus is spoken about within secular and sacred epistemes of infection and riverine pollution; among colonial disease-control agents, historians, biologists, and doctors; and in contemporary museums and discussions about the crisis of antimicrobial resistance (AMR). For recent ethnographic accounts that include the

interactions of humans, microbes, and other plant or animal species, see Guthman 2019 and Wanderer 2020.

11. Murray-Rust et al. 2019.

12. Rao 2018. The human-machine-microbe interfaces are complicated and messy and do not follow "smart" pathways. I have seen new monitoring devices in a few government, hotel, and university wastewater plants and these are supposed to beam up water quality data throughout the day as the wastewater passes out of tertiary treatment. I have seen on many occasions that municipal STPs have these sensors and monitoring devices installed to measure water exiting tertiary treatment phases, but operators do not run them, or they are on and then offline due to power outages or breaks in internet connectivity. These breakages in real-time data collection are also convenient when authorities do not want the treatment plant performance to be revealed to others.

13. Several chapters will explain how decentralized or distributed governance feeds into or enables engaged or "smart" citizenship. Sadoway and Shekhar (2014) have defined smart citizenship as citizen-driven approaches that engage citizens in complementary digitally mediated and face-to-face processes that respect local knowledge systems. Their research has identified how information and communication technologies (ICTs) can serve to spotlight overlooked or undervalued urban infrastructure and planning and environmental issues, such as the need for public toilets, road safety, and pro-pedestrian planning.

14. Flemming, Wingender, and Szewzyk 2016.

15. Larkin 2013.

16. Von Schnitzler 2008.

17. Roy and Ong 2012.

18. Hess and Sovacool 2020; Strauss, Rupp, and Love 2013.

19. Anand 2017; Barnes 2017; Bjorkman 2015; Desai, McFarlane, and Graham 2015; Karpouzoglou and Zimmer 2016; Morales 2016; Workman et al. 2021.

20. Reno 2014, 2020.

21. Schindler and Demaria (2016) have argued that human societies are actual metabolisms whose reproduction is dependent upon the continual use of energy and materials in various forms.

22. See Rath et al. 2020 for the status of FSM in India. For examples of FSM in other countries, see Yesaya and Tilley (2021) on the problems in Malawi.

23. Barr 2019; Singh 2014.

24. See Furlong et al. (2019) for the focus on everyday practices in water research. See also Kemerink-Seyoum et al. (2019) on sociotechnical tinkering in irrigation systems.

25. Communities and businesses use bioreactors with other kinds of devices to filter and purify water. The bioreactors and filtration devices are also connected to larger infrastructures that require pipes and toilets, on one end, and

pipes to drainage and recycling systems on the other. Humans are intimately connected to some sections of these infrastructural components, such as toilets, but may not know much about the other components such as drains and treatment plants. The case studies in this book highlight new forms of learning as communities and businesses experiment with machine building and learn how to visualize the larger infrastructural components that bioreactors must connect to. These experimenters have control over the components on their premises but may have to find ways to link their systems with other grids that are more centralized. Moreover, humans must engage in cleaning, maintenance, and repair practices that endanger their bodies and health. Infrastructures involve humans in tasks that are integral to optimization.

26. For research on decentralized sanitation, see Alexander et al. 2008; Chopra and Das 2019; Frijns et al. 2016; Hurlimann and McKay 2007; Jamwal et al. 2014; Kuttuva, Lele, and Mendez 2018; Molinos-Senante, Hernández-Sancho, and Sala-Garrido 2011; Lienhoop et al. 2014; Ravishankar, Nautiyal, and Seshaiah 2018; Roomratanapun 2001; Suneethi et al. 2015.

27. In the United States, decentralization is enhanced by models of distributed systems in computer science. Distributed systems are computerized networks that work rather independently in different geographical locations but are linked to a central network either physically or by management. The collection of independent computers appears to the users of the system as a single coherent system. Models of distributed systems have also been used in corporate decision-making, city planning, digital communications through the internet, and finance mechanisms such as bitcoin, with the goals of generating self-governance and the horizontal diffusion of authority, trust, and consensus-based decision-making. These models do not propose to replace formal government but act within social, political, and legal structures to advance activities that are beyond the scope or abilities of a centralized agency.

28. Water Environment Federation, Water Science and Engineering Center 2019:3.

29. Liu et al. 2020.

30. Randeria 2003b.

31. Lewis 2013.

32. Roy 2005, 2009, 2012.

33. Mathur 2015.

34. Hull 2012.

35. Joshi and Shambaugh 2018.

36. Murty and Kumar 2011.

37. Olson 1965.

38. Chidambaram 2020:10.

39. Urade and Gondane 2021.

40. Scruggs and Heyne 2021.

1. SANITATION AND INSTITUTIONAL COMPLEXITY

1. Baum, Luh, and Bartram 2013.

2. Lefebvre 2018.

3. Radcliffe and Page 2020; Al-Saidi (2021:3) notes that indirect reuse water is called "blended water" in some places.

4. Dawoud 2017; Zhu and Dou 2018.

5. Alley 2002, 2014, 2015, 2016, Alley and Mehta 2022; Rohilla and Dwivedi 2013; Sanghi 2014; Tare and Roy 2015.

6. Barr (2019:41) writes about the importance of sanitation for women: "In May of 2014, two teenage girls in Badaun, Uttar Pradesh, went out to the field in the night, supposedly to relieve themselves. They were gang raped and found hanging from a mango tree the next morning. The Badaun Rapes, as they came to be known in media shorthand, were vivid and horrifying. The event attracted international attention, with different actors wrestling over the meanings of the event. News reports began to focus on the fact that these girls were ostensibly going outside to relieve themselves, and this act of gendered violence suddenly became about sanitation. Sulabh International, one of the NGOs that was studied for this work, "adopted" the village and built toilets for everyone in the village. Other NGOs and outlets used this incident to underline how sanitation and women's safety are entwined and that sanitation is important (Dash 2014; Jitendra 2015). Throughout media coverage of sanitation and the conversations around it, people continually refer to the importance of sanitation in protecting the safety and dignity of women."

7. Barr 2019; Desai, McFarlane, and Graham 2015.

8. Improved sewage transport, disposal, and treatment have also been supported through the mission called AMRUT or the Atal Mission for Rejuvenation and Urban Transformation and the Smart Cities Program. Altogether, these programs have built the momentum for more inclusive sanitation.

9. Barr (2019:31) notes, "This practice is most commonly recognized as an issue in South Asian countries because of the way it is associated with caste discrimination. Those who are employed to be manual scavengers historically have been predominantly from Dalit castes. They are considered to be ritually polluted by birth and unfit to do any other labor, and society relegates the handling of shit and waste to these groups (Gita 2011; Singh 2014; Sagar 2017). Currently, Dalits seem to dominate in manual scavenging, but migrants (particularly from poorer parts of India) doing this work is also quite common. There is a severe lack of solid data around the issue, however, meaning that exact numbers of manual scavengers are uncertain" (Barr 2019).

10. However, both the Directive Principles and Fundamental Duties are not enforceable by courts unless they are made enforceable by parliamentary law.

11. India Constitution, amended by the Constitution (Forty-Second Amendment) Act, 1976; Bhatt 2004.

12. Bhatt (2004).

13. India Constitution, Art. 48A.

14. India Constitution, Art. 51A(g).

15. The Water (Prevention and Control of Pollution) Act, No. 6 of 1974; India Code (1974).

16. The Water (Prevention and Control of Pollution) Act, No. 6 of 1974; India Code (1974).

17. Rosencranz et al. 1992.

18. The Water (Prevention and Control of Pollution) Act, No. 6 of 1974; India Code (1974) (discussing the Central Pollution Control Board under section 3 and the State Pollution Control Boards under section 4); See also Rosencranz et al., 1992:153n10.

19. See Rosencranz et al. 1992:154n10; see also Singh 1995.

20. Ministry of Environment and Forests 1994.

21. Bhuwania 2017:157.

22. Alley 2009: 799; Bharucha 1998.

23. See Kapur 1998: xi, n19, and Bharucha 1998:vii, n20.

24. *Suo motu* occurs when a court acts on its own interpretation of need or necessity rather than on a problem highlighted by the petitioner. See Razzaque 2004.

25. Ganguli 1998:A-13n18. Ganguli has noted that *suo motu* interventions, the appointment of investigative committees, and monitoring exercises to ensure conformity with the law all signify a transformation in the role of the judge. Justice Kuldip Singh played a charismatic role during his tenure on the Supreme Court bench hearing environmental cases. Over the course of his career, Justice Singh ruled in over one hundred environmental cases, many of them filed by advocate M. C. Mehta.

26. These remedies have been used in a vast array of public interest cases, specifically in Vellore Citizens Welfare Forum v. Union of India and Others, A.I.R. 1996 S.C. 2115; see also Rosencranz et al. 1992:123–30.

27. Jariwala 2000.

28. However, Justice Bhagwati has argued that this power to appoint advisory commissions lies in Order XXVI CPC and Order XLVI of the Supreme Court Rules, 1966, which provide for appointment of commissions for the purpose of examining witnesses, making legal investigations, and examining accounts. See Ganguli 1998:A-14n18.

29. This discussion is supported by a personal assessment of government documents.

30. Alley 2009.

31. Amirante 2012; Kumar 2016; Shrotriya 2015.

32. Amirante 2012.

33. Gill 2017, 2018:17.

34. Gill 2018:19.

35. Manoj Mishra vs. Union of India and Others. Original Application No. 6 of 2012 and M.A. numbers 967/2013 and 275/2014. Judgment dated January 13, 2015.

36. Since 2010, the Central Ground Water Authority (CGWA) has assessed groundwater resources in units, as blocks, talukas, mandals, and watersheds. These assessment units are based on two criteria: (a) stage of groundwater development, and (b) long-term pre- and post-monsoon water levels. They compute long-term groundwater level trends over a ten-year period. They use four categories to allow extraction or set limits: (1) "Safe" areas which have groundwater potential for development; (2) "Semi-critical" areas where cautious groundwater development is recommended; (3) "Critical" areas; and (4) "Over-Exploited" areas, where there should be intensive monitoring, evaluation, and water conservation measures. Of the 6,607 assessment units (blocks, mandals, talukas, and districts) across the country, 1,071 are overexploited, 217 are critical, 697 are semi-critical, 4,580 are safe, and 92 are saline.

37. The Central Ground Water Authority was constituted under Section 3 (3) of the Environment (Protection) Act, 1986 to regulate and control development and management of groundwater resources in the country.

38. Cases focusing on pollution of the River Ganga and its tributaries are many, and occasionally the justices in the NGT, like those in the Supreme Court, will combine or "club" cases with similar pleas and purviews to hear them as a collective.

39. Order dated July 13, 2017, M. C. Mehta vs. Union of India. Original Application No. 200 of 2014 (C. Writ Petition No. 3727/1985) (M.A. No. 594/2017 & 598/2017), 82–83.

40. The responding agencies were bound to comply at least in performance. However, there have been many acts of noncompliance and the NGT has been forced to follow up and monitor activities and compliance with previous orders of its court and of the High Courts and Supreme Court. By 2018, the NGT had issued over one hundred orders and judgments in the Ganga case regarding wastewater treatment plants. The orders and judgments doled out punishments to polluters; closed industries operating without pollution-control facilities; and required hotels, ashrams, and housing complexes to install their own wastewater treatment plants.

41. Order dated October 16, 2017, M. C. Mehta vs. Union of India. Original Application No. 200 of 2014 (C. Writ Petition No. 3727/1985) (M.A. No. 594/2017 & 598/2017).

42. Hess 2014.

43. I recorded this statement in my notes taken during the session of the day.

44. I recorded these statements in my notes taken during the session of the day.

45. Since I also knew about this initiative through my own networks, I figured it was common knowledge among those arguing and hearing the Ganga case.

46. A Notification is issued by a central or state government to exercise the power of a legislative enactment (Parliamentary/Assembly). These powers are in abundance in taxation (direct/indirect) legislations. These notifications generally lay down the law taking care of some procedural aspects of the enactment. Notifications are published in the Official Gazette.

47. Roy (2005, 2009, 2012).

48. These orders were stayed in the Supreme Court fairly quickly, sending a message to the lower courts that such sweeping regulatory measures were not to be undertaken by a lower court. On rights of nature cases, see Alley 2019, Alley and Mehta 2022, O'Donnell 2018 and O'Donnell and Talbot-Jones 2018.

49. W.P.PIL No. 140 of 2015, Lalit Miglani v. State of Uttarakhand

50. CPCB 2015. Directions under Section 18(1)(b) of the Water (Prevention and Control of Pollution) Act, 1974, regarding treatment and utilization of sewage. April 21.

51. Forest, Ecology and Environment Secretariat Notification No. Fee 316 EPC 2015 Bengaluru, dated January 19, 2016.

52. Forest, Ecology and Environment Secretariat Notification No. Fee 316 EPC 2015 Bengaluru, dated January 19, 2016.

53. Evans, Varma, and Krishnamurthy 2014:3.

54. Notification No. BWSSB/C/CAO-S/5008/2017-18, Bengaluru, dated: February 21, 2018. The Bangalore Sewerage (Amendment) Regulations, 2018.

55. Chauhan et al. 2016:8.

56. Report of Joint Inspection Team regarding Hotels in NDMC Area with respect to Hon'ble NGT orders dated 15th, 17th and 24th August, 2017 in the matter of Almitra H. Patel & Anr. Vs. UOI & Others. and various connected matters. Original Application No. 291/2017 (Earlier O.A. No. 199/2014); Original Application No. 293/2017(Earlier O.A. No. 199/2014); Original Application No. 294/2017 (Earlier O.A. No. 199/2014); Original Application No. 295/2017 (Earlier O.A. No. 199/2014); Original Application No. 298/2017 (Earlier O.A. No. 199/2014); Original Application No. 301/2017 (Earlier O.A. No. 199/2014); Original Application No. 307/2017 (Earlier O.A. No. 199/2014); Original Application No. 308/2017 (Earlier O.A. No. 199/2014); Original Application No. 311/2017 (Earlier O.A. No. 199/2014); Original Application No. 349/2017 (Earlier O.A. No. 199/2014).

57. Shubhra Pant, How Much Sewage Are Housing Societies Treating? GMDA. *Times of India*, November 19, 2021, https://timesofindia.indiatimes.com /city/gurgaon/gmda-how-much-sewage-are-hsg-societies-treating/articleshow /79291698.cms.

58. Ravi Diwaker, "Noida: Housing Societies to Face Action for Violating STP Norms," *Magicbricks*, February 26, 2021, https://content.magicbricks.com /property-news/delhi-ncr-real-estate-news/noida-housing-societies-to-face -action-for-violating-stp-norms/119507.html.

2. INVENTING BIOREACTORS

1. Alley, Maurya, and Das 2018.

2. Maurya et al. (2017:3) have related: "The 2015 CPCB report *Inventorization of Sewage Treatment Plants* states that there are a total of 816 municipal sewage treatment plants (STPs) across India, but of these, 522 are operational, 79 plants are non-operational, 145 are under construction, and 70 are proposed. If all 816 STPs operate at full capacity, only 23,277 million litres per day (MLD) or 38% of the total sewage load would be treated. The total sewage generated across India is over 61,754 MLD."

3. Interview with a chief engineer at the city of Chennai wastewater treatment plant, October 2017.

4. See Scruggs et al. (2020) on the health implications from toxicity during direct potable reuse.

5. Auroville has a special status as a foundation. Legally, they fall under the Ministry of Human Resources. The legal umbrella was granted by an Act of Parliament in 1988, and that status facilitated some of their work but has made other things difficult. It gives a legal structure to the whole experimentation process. They get local approvals and use a tender and it is channeled through state officials. But they have the ability to avoid the corruption of the state. Auroville has a governing body that takes care of their activities. Residents run all their activities and they are responsible for them. They submit their annual report to the Government of India and a seven-member board audits their accounts. They have an international connection with UNESCO, which recognizes their work for humanity.

6. They make their money with the vortex, which they started selling in 2011. In 2017, the prefabricated costs were around 35 lakh rupees (3,500,000) for an 85 kld plant. The maintenance is focused on the pumps. They give a seven-year guarantee that pumps will work but they will have to be replaced at some point. For a larger unit, the annual maintenance costs are 3–4 lakh rupees (300,000–400,000 rupees) for manpower, desludging, and cleaning but not including pump replacements. By contrast, he noted that a conventional STP of the same size would cost 80 lakh rupees (8,000,000) for annual maintenance, not counting replacements. Some revenue can be generated by the STP. For instance, the eye hospital in Chennai generates too much treated wastewater, so the hospital sells the remaining water to a tanker company.

7. Interview with a principal inventor of the SBT technology.

8. For nitrogen fixation as well, although this is debatable, he adds. His method can avoid nitrogen poisoning.

9. Their main office is in Bangalore, and they have satellite offices in Nagpur, Jaipur, and Jharkhand. They have a representative in Chennai and a fecal sludge management (FSM) site in Leh, the capital of Ladakh.

10. Interview with director of Ecoparadigm, who was a founding member of CDD.

11. Interview with high-ranking engineer at CDD.

12. Interview with director of Ecoparadigm.

3. DOUBLE BURDENS

1. For work on water provisioning in low-income communities and disproportionate burdens on women, see Bapat and Agarwal (2003) and Truelove (2011). See McGranahan (2015) on challenges in organizing sanitation in low-income communities and especially on the problem of affordability.

2. Since 2015, builders have had to install the STP at the time of construction and they have created something that looks like a treatment plant. However, this is sometimes just a single tank that does not operate properly as a treatment unit. After construction, some builders and contractors operate and monitor the wastewater treatment plants in the first year or two and then hand them over to the resident societies for onward operation and maintenance.

3. Alexander et al. 2008; Chopra and Das 2019; Frijns et al. 2016; Hurlimann and McKay 2007; Jamwal et al. 2014; Kuttuva, Lele, and Mendez 2018; Molinos-Senante, Hernández-Sancho, and Sala-Garrido 2011; Lienhoop et al. 2014; Ravishankar, Nautiyal, and Seshaiah 2018; Roomratanapun 2001; Sengupta 2018; Suneethi et al. 2015.

4. Biswas and Jamwal 2017.

5. In Bangalore, the water table lies at a depth of eight hundred feet below the surface. According to Goldman and Narayan (2019), in 2015 there were 105,500 private bore wells registered with the Bangalore Water Supply and Sewerage Board and over 200,000 unregistered bore wells. Across Karnataka, 5,000 new bore wells are sunk every month, in many cases to replace ones that have dried up. Other cities have groundwater tables lurking at two hundred to three hundred feet below the surface. In some places, groundwater is contaminated by industrial activities and rendered unusable. Groundwater in the peri-urban areas of Kanpur is severely contaminated by tanneries and other industries. Residents of the largest resettlement village outside Delhi are not able to use their groundwater for potable and bathing uses and must limit use to washing clothes. They rely upon water tankers from the Delhi Water Board for their drinking-water supply. These are just a few of the many examples of the stresses related to groundwater accessibility and usability.

6. The BWSSB provides around 1,170 mld of piped water to Bangalore, which is derived from surface water (1,120 mld) and groundwater (70 mld). However, the 8.8 million people that live in Bangalore Metropolitan Area (BMA) are expected to have a water demand of 2,550 mld by 2036, with predicted shortfalls

of 220 mld and 1,050 mld respectively (Kelkar and Thippeswamy, 2012; Kelkar, Wable, and Shukla 2012). The actual quantity of water used is higher due to private supplies. The quantity of wastewater generated is approximately 1,200 mld of which 120–350 mld is treated (Kelkar and Thippeswamy 2012). Around 45 percent of households in Bangalore are connected to sewers, with the remainder having septic tanks, pit latrines, or disposing directly to the environment (Census 2011).

7. These are all larger STPs with the smallest under the BWSSB at 1.5 mld. This information comes from interviews with key actors in the STP scape.

8. Interviews with officials in BWSSB, CDD, and Ecoparadigm.

9. Krishnamurthy (2019) explained, "A faecal sludge treatment plant is housed in Devanahalli. To establish this plant, the CDD partnered with Devanahalli's Town Municipal Cooperation. Here they co-compost treated faecal sludge with municipal wet waste, which, according to their laboratory tests, keeps pathogens out of the final manure. They also find that the co-composted manure improves the water-holding capacity of the soil and increases crop yield. The co-composted manure, the CDD says, has seen demand from the farmers in Devanahalli."

10. Olson 1965; Chidambaram 2020; McGranahan 2015.

11. Olson 1965.

12. Many communities in Bengaluru have their own borewells and the estimate is that there are fifty thousand private borewells in Bangalore. Now residents need permission from the municipality, the BBMP, to drill for a new borewell but the monitoring is rather weak.

13. See Ho's (2019) comparison of India and China on social contracts for public goods.

14. According to the CDD, there were 120 houses connected to the STP in this colony.

15. This is a general problem with 'willingness to pay' questions, since they anticipate conditions that residents are not yet experiencing.

16. Auerbach 2020.

17. See CURE 2007.

18. CURE 2007.

19. CURE 2007, 34.

20. Maurya et al. 2017.

21. Chidambaram 2020.

22. McGranahan (2015:245–46) explains Ostrom's argument that the sharp conceptual divide between government and civil society is a trap, hiding the potential synergies that can be gained from coproducing goods and services. Defining coproduction as "a process through which inputs from individuals who are not in the same organization are transformed into goods and services," Ostrom (1996:1083) concluded that "co-production of many goods and services normally considered to be public goods by government agencies and citizens organized

into polycentric systems is crucial for achieving higher levels of welfare in developing countries, particularly for those who are poor." McGranahan argues, "There are many reasons why sanitary improvements are often best co-produced, particularly in informal urban settlements. Some of the reasons are related to the sorts of incentive problems and competencies involved in collective action at different scales (community and city). But most important from the perspective of advocates of coproduction, by co-producing sanitation residents of informal settlements should be able to secure better services from their governments, and in return public agencies should be able to secure more public-spirited behavior from some of their worst-off citizens." Drawing from these perspectives, I emphasize that the community needs to be the driver of the decisions and needs a benefit from decentralized wastewater treatment to justify the ongoing expense.

4. HORTICULTURAL, PARTIAL, AND OFF-GRID REUSE

1. Starkl, Anthony, et al. 2018.

2. Starkl, Anthony, et al. 2018.

3. Directions of NGT order dated June 11, 2015, in the matter of OA No. 6/2012 & 300/2013, accessed July 14, 2018, http://delhi.gov.in/wps/wcm/connect/07be3 30048dbd704b6f9ff7a2b587979/Directions_Clarifications_NGT_11.6.2015.pdf ?MOD=AJPERES&lmod=-287594179.

4. Pandey 2015

5. Delhi Jal Board Water Tariff revised February 1, 2018.

6. The Okhla STP is one of the largest in Delhi and is located at the downstream end of the Yamuna River. They send the treated water to the city through a number of old pipelines laid decades ago by the Central Public Works Department.

7. Report of Effluent Usage at STPs, Delhi Jal Board, unpublished document. September 2017.

8. Lloyd Owen 2016.

9. The CSIR-National Environmental Engineering Research Institute (CSIR-NEERI) is a research institute created and funded by the Government of India.

10. The International Water Management Institute (IWMI) and Biome Environmental Trust describe a more encouraging pathway for the fecal sludge: "Faecal sludge evacuation service providers, known locally as 'honey suckers' are invaluable, meeting the needs of the population who are not served by the government sewage system. They often dispose of the FS [fecal sludge] on agricultural land, at the request of the farmers, thereby helping to close the nutrient loop. There are also many companies engaging in composting (at various scales), biogas generation and wastewater treatment or related services. Like FS use, wastewater use has been observed to take place informally, for instance, when

farmers tap into untreated wastewater flows or fishing takes place in lakes receiving wastewater," 3; "There were an estimated 300 honeysuckers in 2012 but numbers appear to be growing," 71. See also 156–59. Resource Recovery and Reuse (RRR) Project (2012).

11. Evans (2009:3) stated, "The KSPCB's rules are enforced and monitored through a system of applications for consent, which requires construction companies to apply for consent for establishment (CFE) prior to building a property, followed by a consent for operation (CFO) prior to commissioning the STP. The CFE must include plans and specifications for treatment and the CFO is used to check that the builders have implemented the STP as proposed. Consent must be obtained annually and STP owners must provide evidence that their STP is operating successfully and meeting the discharge standards (KSPCB, 2004). More recently, the KSPCB has released urban reuse norms, set from a human health perspective that apply to domestic sewage treatment, treated wastewater quality and wastewater discharge in new real estate developments (Table 3). These make the reuse of treated water for non-potable purposes within apartments and commercial buildings mandatory and stipulate that there should be zero disposal (KSPCB Memorandum No. 3080, Dated: 16.8.2012)."

12. Interview with resident.

13. Blair (2018:14) notes, "Residential Welfare Associations (RWAs) have developed as intermediary CSOs [civil society organization] between citizen and state in a number of Indian cities over the last several decades. In many cases, the relationship has become a partnership (*bhagidari* in Hindi) in which the RWA and the municipal government jointly plan and even implement public service delivery (e.g., sanitation, electricity, water). Sometimes an RWA will actually provide the service itself (e.g., security). In some cases, the partnership begins with an initiative from the state side, while in others a CSO persuades the state to engage with it."

14. Dual plumbing was mandated for new housing complexes starting in 2015, see directions issued by the Karnataka State Pollution Control Board No. PCB/074/STP/2012 dated December 5, 2015. These were copied from the Central Pollution Control Board's directions that had been issued earlier in 2015. The directions set the standards for key water-quality parameters, and set the BOD in the STP outfluent at 10 mg/l.

15. McDonald 2018.

16. So far, the cases in this book have shown that the terrain of action involves communities or businesses as well as the public or private sectors. In these public-private partnerships, the arrangements are more trilateral, with communities, businesses, or institutions in partnership with public agencies and private companies. The public agencies are the regulators, and the private companies are the builders and operators, or those conducting maintenance and repair. The communities, businesses, and institutions form the centers of financial decision-making and become the portals through which funds are raised and assigned.

17. RWAs have submitted writ petitions against slum habitations near their complexes; they have protested unauthorized dwellings and sites of religious worship on the basis of property rights. See Blair 2018; Coelho and Venkat 2009; Harriss 2007; Lemanski and Lama-Rewal 2013.

18. Our survey asked whether there was a subsidy for setting up STPs in housing communities and no one mentioned one.

19. Although the builder's role in initial failures of decentralized STPs is well known among housing society residents, there is surprisingly little literature on the subject.

5. CLOSED LOOPS AND EMERGING REUSE

1. Isenhour and Reno 2019; Fletcher and Rammelt 2017.

2. Central Pollution Control Board 2015.

3. Grönwall and Jonsson 2017:17.

4. Grönwall and Jonsson 2017:13.

5. M. C. Mehta versus Union of India and Original Application No. 501 of 2014 (M.A. No. 404 of 2015) Anil Kumar Singhal versus Union of India & Others and Original Application No. 146 of 2015 Society for Protection of Environment & Biodiversity & Anr. versus Union of India & Others and Appeal No. 63 of 2015 Confederation of Delhi Industries & CEPT Societies (An Organisation of CETP Societies) versus D.P.C.C. & Others and Original Application No. 127 of 2017 J. K. Srivastava versus Central Pollution Control Board & Others and Original Application No. 133/2017 (Writ Petition (C) No. 200/2013) Swami Gyan Swarop Sanand versus Ministry of Home Affairs & Others Judgment 13th July, 2017.

6. M. C. Mehta versus Union of India 2014 and Others.

7. M. C. Mehta versus Union of India 2014 and Others.

8. Forest, Ecology and Environment Secretariat Notification No. FEE 316 EPC 2015, Bengaluru, India, January 19, 2016.

9. Interview with leading engineering professor, IIT-M, October 5, 2017.

10. Krishna Chaitanya and Krishna 2017.

11. Krishna Chaitanya and Krishna 2017. The following description of the project on the website for the Saraswati 2 project, a project combining experiments across many universities in India, shows how this system, like many others, is being continuously evaluated and optimized to meet reuse requirements. The web description states: "The SBR plant at IIT Madras features a tertiary treatment system consisting of ultra-filtration and ozonation. Although an anoxic system is provided in the SBR, the treated water still has significant levels of nitrate (10–25 ppm) and nitrite (7–9 ppm). As a result of residual nutrients, the storage ponds have a significant amount of eutrophication. This also limits the reuse potential of the treated water. Providing a treatment system to reduce the residual nutrients will increase the treated water reusability significantly. The benefits of

the technology: Providing a treatment system that reduces the residual nitrogen-based nutrients significantly increases usability of the treated secondary and tertiary effluents. The main advantages of the technology include: Bioreactor kept separate from the water treated—so no need to remove back-contamination of bacteria or organic load; Simple to operate—just two process flow streams and simple controls for pH; and it is flexible in that multiple modules in parallel allow increase in capacity and multiple modules in series allows to reach whatever extent of nitrate/nitrite removal that is required." See https://projectsaraswati2 .com/pilot-sites/pilot-10-nitrate-removal-using-iemb-reactor/.

12. Interview with manager of Engineering Department, Marriott Renaissance Hotel and Conference Center, October 14, 2017.

13. World Bank 2017

14. Report of Joint Inspection Team regarding Hotels in NDMC Area with respect to Hon'ble NGT orders dated 15th, 17th and 24th August 2017 in the matter of Almitra H. Patel & Anr. vs. UOI & Others and various connected matters. Original Application No. 291/2017 (Earlier O.A. No. 199/2014); Original Application No. 293/2017 (Earlier O.A. No. 199/2014); Original Application No. 294/2017 (Earlier O.A. No. 199/2014); Original Application No. 295/2017 (Earlier O.A. No. 199/2014); Original Application No. 298/2017 (Earlier O.A. No. 199/2014); Original Application No. 301/2017 (Earlier O.A. No. 199/2014); Original Application No. 307/2017 (Earlier O.A. No. 199/2014); Original Application No. 308/2017 (Earlier O.A. No. 199/2014); Original Application No. 311/2017 (Earlier O.A. No. 199/2014); Original Application No. 349/2017 (Earlier O.A. No. 199/2014).

15. This is a qualitative statement that I developed from review of the interview and survey data, from visits to many housing complexes in the region and from other documentary information.

16. In 2010, the Punjab and Haryana High Court had directed the district administration to ban groundwater use for construction and other uses that were not for domestic consumption. See Arora (2019).

17. It is important to note that using water for construction is a significant draw on supplies. That builder used about fifty to sixty kiloliters of water per day for their on-site uses. They also used water from private tankers during the construction. *Down to Earth* reported in 2012: "The Punjab and Haryana High Court has directed the principal secretary of Haryana's town and country planning department to ensure that no tube well is operational for construction purposes in Gurgaon. The official would be personally liable if the directives are not implemented in earnest, the court said on August 21. The order follows an earlier order of the court prohibiting groundwater extraction for building projects. To help out the developers, the state authorities have, meanwhile, made available some of the city's treated sewage so that construction activity can restart." Seth 2012.

18. If there had been more transparency in the system, I could have figured out where the tanker water went. However, on that day, the tanker driver on-site

was not at all willing to talk to me and I knew it would take more time to investigate where they actually take the water that the RWA pays them to take away.

19. In 2011, an expert in sewage treatment noted that, "Our experience from operating STPs in over fifteen high-rise complexes with number of units varying from 150 to 1,200 flats, indicate[s] that the approximate percentage of treated water required to be disposed off varies between 25 to 40%." CommonFloor Editorial Team 2011.

6. PRETEND MACHINES

1. Hull 2015; Mathur 2015.

2. Blair 2018; Brunner et al. 2014; Ghosh 2019:218; Sharma, Yadav, and Gupta 2016.

3. Blair 2018; Halley and Shore 2005; Muir and Gupta 2018; Sharma 2018.

4. Bourdieu's (1984) notion of habitus could be invoked here to describe the interrelations between actors' involvements with machinery and their positions of intention within the structures of sanitation technologies. See also the argument by Furlong et al. (2019) that focuses on everyday practices in water research and Kemerink-Seyoum et al. (2019) on sociotechnical tinkering.

5. Starkl, Brunner, and Stenströme 2017.

6. Starkl et al. 2017:133.

7. Uttar Pradesh Water Supply and Sewerage Act, 1976.

8. The Uttarakhand High Court order detailed all the drains emptying into the Ganga in Uttarakhand. In this document, the statuses of all the STPs in the state were also detailed. See in the High Court of Uttarakhand at Nainital Writ Petition (PIL) No. 49 of 2018, "In the matter of Suo Moto Cognizance Regarding Contamination of Water of River Ganga vs. State of Uttarakhand and others." For an audit of Pey Jal Nigam facilities, see *Report of the Comptroller and Auditor General of India for the Year Ended 31 March 2017*. Government of Uttarakhand Report No. 1 of the year 2018. See also Singh et al. 2016 for a CPCB report on water and sewage for ashrams and hotels in Rishikesh.

9. This is similar to monitoring in Bengaluru, where, in discussions with the RWAs and the KSPCB, Kuttuva, Lele, and Mendez (2018:16) found that enforcement by KSPCB officers targeted the bigger apartment complexes.

10. Amirante 2012; Bhuwania 2017; Kumar 2016.

11. M. C. Mehta v. Union of India, C. Writ Petition No. 3727/1985, National Green Tribunal.

12. These are reported a year later in the Report of the Comptroller and Auditor General of India for the year ended March 31, 2017.

13. Order dated December 10, 2015, in M. C. Mehta v. Union of India, C. Writ Petition No. 3727/1985, National Green Tribunal.

14. Order dated July 13, 2017, in M. C. Mehta v. Union of India, C. Writ Petition No. 3727/1985, National Green Tribunal.

15. See the Government of India's Jal Shakti campaign at: https://ejalshakti .gov.in/JSA/JSA/Home.aspx.

16. The survey of business and resident perspectives on wastewater and reuse showed significantly less interest in treatment and reuse in Rishikesh, when compared with responses from the National Capital Region and Bangalore.

17. Kuttuva et al. 2018.

18. Kuttuva et al. 2018:5.

19. Copeman and Ikegame 2012; Ludden 1996; van der Veer 1994.

7. CONCLUSIONS

1. Many studies advocate for community-based water management, but in strong centralized systems it is hard to find the legitimate avenues for communities to act alone. The cases in this book have shown multiple unintended points of entry because the centralized, or on-grid supplies of water are woefully inadequate to meet residents' needs, especially in peri-urban areas.

2. Gunderson and Holling 2002.

3. Pierce and Gonzalez 2017; Brunner et al. 2018.

4. Evans, Varma, and Krishnamurthy 2014:4.

5. Kundu 2011.

6. Srivastava (2021) has described how the Central Public Sector Undertakings (CPSUs) have played a vital role in the development of India's economy. These CPSUs grew from 5 to 348 and increased investment from Rs. 29 crore to Rs. 16.4 trillion by the end of 2018–19. The CPSUs have played a strategic role in achieving higher economic growth, self-sufficiency in production of goods and services, and low, stable prices.

7. See Dasgupta (2015) on middle-class and IT professionals and their interactions with the state and NGOs in water supply. She refers to this movement as the "middle-class capture of the state."

8. Harriss 2007:2721.

9. On limitations in fecal sludge management, see Mallory et al. 2020.

10. Roller and Mayorga 2017; see also Scruggs and Heyne 2021

11. Wright, Godfrey, and Armiento 2019.

12. Arora et al. 2015:632; Hong et al. 2013; Lamba and Ahammad 2017. Among the new studies advocating surveillance of sewage for early warning of COVID cases, Peccia et al. (2020) argue that raw wastewater and sludge-based surveillance is particularly useful for low- and middle-income countries where clinical testing capacity is limited. See also Gahlot et al. 2023.

References

Alexander, K. S., J. C. Price, A. L. Browne, Z. Leviston, B. J. Bishop, and B. E. Nancarrow. 2008. *Community Perceptions of Risk, Trust and Fairness in Relation to the Indirect Potable Use of Purified Recycled Water in South East Queensland: A Scoping Report.* Technical Report No. 2. Queensland, Australia: Urban Water Security Research Alliance.

Alley, Kelly D. 2002. *On the Banks of the Ganga: When Wastewater Meets a Sacred River.* Ann Arbor: University of Michigan Press.

———. 2009. "Legal Activism and River Pollution in India." *Geo. Int'l Envtl. L. Rev.* 21: 793.

———. 2014. "The Developments, Policies and Assessments of Hydropower in the Ganga River Basin." In *Our National River Ganga: Lifeline of Millions.* Rashmi Sanghi, ed., 285–305. Cham, Switzerland: Springer.

———. 2015. "Killing a River: The Failure of Regulation." In *Living Rivers, Dying Rivers: A Quest through India.* Ramaswamy Iyer, ed. Delhi: Oxford University Press.

———. 2016. "Rejuvenating Ganga: Challenges and Opportunities in Institutions, Technologies and Governance." *Tekton: A Journal of Architecture, Urban Design and Planning* 3(1).

———. 2019. "River Goddesses, Personhood and Rights of Nature: Implications for Spiritual Ecology." *Religions* 10(9): 502. https://doi.org/10.3390/rel10090502.

Alley, Kelly D., Jennifer Barr, and Tarini Mehta. 2018. "Infrastructure Disarray in the Clean Ganga and Clean India Campaigns." *WIREs Water* 5: e1310.

Alley, K. D., N. Maurya, and S. Das. 2018. "Parameters of Successful Wastewater Reuse in Urban India." *Indian Politics and Policy* 1(2).

Alley, Kelly D., and Tarini Mehta. 2022. "Contradictions in Pollution Control: Religion, Courts and the State in India." In *Climate Politics and the Power of Religion*, E. Berry, ed., 147–75. Bloomington: Indiana University Press.

Al-Saidi M. 2021. "From Acceptance Snapshots to the Social Acceptability Process: Structuring Knowledge on Attitudes Towards Water Reuse." *Front. Environ. Sci.* 9: 633841. doi: 10.3389/fenvs.2021.633841.

Amirante, D. 2012. "Environmental Courts in Comparative Perspective: Preliminary Reflections on the National Green Tribunal of India." *Pace Environmental Law Review* 29(2): 441–69.

Anand, Nikhil. 2017. Hydraulic City: Water and the Infrastructures of Citizenship in Mumbai. Durham: Duke University Press.

Arora, M., H. Malano, B. Davidson, R. Nelson, and B. George. 2015. "Interactions between Centralized and Decentralized Water Systems in Urban Context: A Review." *Wiley Interdisciplinary Reviews: Water* 2(6): 623–34.

Arora, Shilpy. 2019. "File Reports on Groundwater Extraction in Gurugram, Govts Told." *Times of India*, April 19. http://timesofindia.indiatimes.com /articleshow/68961325.cms?utm_source=contentofinterest&utm_medium= text&utm_campaign=cppst.

Auerbach, Adam Michael. 2020. *Demanding Development: The Politics of Public Goods Provision in India's Urban Slums*. Cambridge: Cambridge University Press.

Bapat, M., and I. Agarwal. 2003. "Our Needs, Our Priorities; Women and Men from the Slums in Mumbai and Pune Talk about Their Needs for Water and Sanitation." *Environment and Urbanization* 15(2): 71–86.

Barnes, Jessica. 2014. "Mixing Waters: The Reuse of Agricultural Drainage Water in Egypt." *Geoforum* 57: 181–91.

———. 2017. "States of Maintenance: Power, Politics, and Egypt's Irrigation Infrastructure." *Environment and Planning D: Society and Space* 35(1): 146–64.

Barr, Jennifer. 2019. "Private Acts, Public Stories: Sanitation NGOs during the 'Clean India' Mission." PhD diss., Emory University.

Baum, R., J. Luh, and J. Bartram. 2013. "Sanitation: A Global Estimate of Sewerage Connections without Treatment and the Resulting Impact on MDG Progress." *Environmental Science & Technology* 47(4): 1994–2000.

Bharucha, S. P. 1998. "Golden Jubilee Year of the Constitution of India and Fundamental Rights." In *Supreme Court on Public Interest Litigation VII*, J. Kapur, ed. New Delhi: LIPS Publications.

Bhatt, S. 2004. *Environment Protection and Sustainable Development*. New Delhi: S.B. Nangia.

Bhuwania, Anuj. 2017. *Courting the People: Public Interest Litigation in Post-Emergency India*. Cambridge: Cambridge University Press.

Biswas, D., and P. Jamwal. 2017. "Swachh Bharat Mission: Groundwater Contamination in Peri-urban India." *Economic and Political Weekly* 52(20): 18–20.

Bjorkman, Lisa. 2015. *Pipe Politics. Contested Water: Embedded Infrastructures of Millennial Mumbai*. Durham, NC: Duke University Press.

Blair, Harry. 2018. "Citizen Participation and Political Accountability for Public Service Delivery in India: Remapping the World Bank's Routes." *Journal of South Asian Development* 13(1): 1–28.

Bourdieu P. 1984. *Distinction: A Social Critique of the Judgement of Taste*. Cambridge, MA: Harvard University Press.

Braun, B. 2007. "Biopolitics and the Molecularization of Life." *Cultural Geographies* 14(1): 6–28.

Brunner, N., M. Starkl, P. Sakthivel, L. Elango, S. Amirthalingam, C. E. Pratap, M. Thirunavukkarasu, and S. Parimalarenganayaki. 2014. "Policy Preferences about Managed Aquifer Recharge for Securing Sustainable Water Supply to Chennai City, India." *Water* 6: 3739–57.

Brunner, N., M. Starkl, A. Kazmi, A. Real, N. Jain, and V. Mishra. 2018. "Affordability of Decentralized Wastewater Systems: A Case Study in Integrated Planning from India." *Water* 10(11): 1644.

Cavanagh, Connor J. 2014. "Biopolitics, Environmental Change, and Development Studies." *Forum for Development Studies* 41(2): 273–94.

Census. 2011. Primary Census Abstracts. Registrar General of India, Ministry of Home Affairs. New Delhi: Government of India.

Central Pollution Control Board. 2015. "Guidelines on Techno-economic Feasibility of Implementation of Zero Liquid Discharge (ZLD) for Water Polluting Industries." New Delhi: Government of India.

Chauhan, S., O. Jensen, G. Sengaiah, and N. Sreenivas. 2016. *Closing the Water Loop: Reuse of Treated Wastewater in Urban India*. https://www.pwc.in /assets/pdfs/publications/2016/pwc-closing-the-water-loop-reuse-of-treated -wastewater-in-urban-india.pdf.

Chidambaram, Soundarya. 2020. "How Do Institutions and Infrastructure Affect Mobilization around Public Toilets vs. Piped Water? Examining Intra-slum Patterns of Collective Action in Delhi, India." *World Development* 132: 104984.

Chopra, V., and S. Das. 2019. "Estimating Willingness to Pay for Wastewater Treatment in New Delhi: Contingent Valuation Approach." *Ecology, Economy and Society–the INSEE Journal* 2(2): 75–108.

Coelho, K., and Venkat, T. 2009. The Politics of Civil Society: Neighbourhood Associationism in Chennai. *Economic and Political Weekly* 44(26/27): 358–67.

CommonFloor Editorial Team. 2011. "STP: KSPCB Norms Puts Apartment Communities in Dilemma." *CommonFloor*, April 13. https://www.common floor.com/guide/stp-kspcb-norms-puts-apartment-communities-in -dilemma-5195.

Copeman, J., and A. Ikegame. 2012. *The Guru in South Asia: New Interdisciplinary Perspectives*. London: Routledge.

CURE. 2007. *Crosscutting Agra Program, Final Report*. August. https://ghn .globalheritagefund.com/uploads/documents/document_2167.pdf.

Dasgupta, Simanti. 2015. *Bits of Belonging: Information Technology, Water, and Neoliberal Governances in India*. Philadelphia: Temple University Press.

Dash, Dipak Kumar. 2014. "Badaun Gang Rape and Murder Retrains Focus on Lack of Sanitation Facilities." *Times of India*, June 2. https://timesofindia .indiatimes.com/india/Badaun-gang-rape-and-murder-retrains-focus-on -lack-of-sanitation-facilities/articleshow/35924064.cms.

Dawoud, M. 2017. "Treated Wastewater Reuse for Food Production in Arab Region." *Arab Water Council Journal* 8(1): 55–86.

Desai, R., C. McFarlane, and S. Graham. 2015. "The Politics of Open Defecation: Informality, Body, and Infrastructure in Mumbai." *Antipode* 47(1): 98–120. doi: 10.1111/anti.12117.

Evans, A. E. V., S. Varma, and A. Krishnamurthy. 2014. "Formal Approaches to Wastewater Reuse in Bangalore, India. Sustainable Water and Sanitation Services for All in a Fast Changing World." Presented at 37th WEDC International Conference, Hanoi, Vietnam.

Fielding, Kelly D., Sara Dolnicar, and Tracy Schultz. 2018. "Public Acceptance of Recycled Water." *International Journal of Water Resources Development* 34(4): 551–86. doi:10.1080/07900627.2017.1419125.

Flemming, H., J. Wingender, U. Szewzyk, et al. 2016. Biofilms: An Emergent Form of Bacterial Life. *Nat Rev Microbiol* 14: 563–75.

Fletcher, R., and C. Rammelt. 2017. Decoupling: A Key Fantasy of the Post-2015 Sustainable Development Agenda. *Globalizations* 14: 450–67. https://doi.org /10.1080/14747731.2016.1263077.

Frijns, Jos, Heather M. Smith, Stijn Brouwer, Kenisha Garnett, Richard Elelman, and Paul Jeffrey. 2016. "How Governance Regimes Shape the Implementation of Water Reuse Schemes." *Water* 8, no. 12: 605.

Furlong, K., D. Roca-Servat, T. Acevedo-Guerrero, and M. Botero-Mesa. 2019. "Everyday Practices, Everyday Water: From Foucault to Rivera-Cusicanqui (with a Few Stops in Between)." *Water* 11(10): 2046.

Gahlot, P., K. D. Alley, S. Arora, S. Das, A. Nag, V. K. Tyagi. 2023. Wastewater Surveillance Could Serve as a Pandemic Early Warning System for COVID-19 and Beyond. *Wiley Interdisciplinary Reviews: Water*: 10(4) July/August. https://doi.org/10.1002/wat2.1650.

Ganguli, A. K. 1998. "In Public Interest: A Review of PIL in the Supreme Court." In *Supreme Court on Public Interest Litigation A-1.*, J. Kapur, ed., 798. Delhi: LIPS.

Ghosh, S. 2019. "The Environment." In *Regulation in India: Design, Capacity, Performance*, D. Kapur and M. Khosla, eds., 203–28. Delhi: Hart.

Gill, G. Nain. 2017. *Environmental Justice in India: The National Green Tribunal*. Abingdon, UK: Routledge.

———. 2018. "Mapping the Power Struggles of the National Green Tribunal of India: The Rise and Fall?" *Asian Journal of Law and Society* 7(1): 85–126. https:doi.org/10.1017/als.2018.28.

Gita, Ramaswamy. 2011. *India Stinking: Manual Scavengers in Andhra Pradesh*. New Delhi: Navayana Publishers.

Goldman, M., and D. Narayan. 2019. Water Crisis through the Analytic of Urban Transformation: An Analysis of Bangalore's Hydrosocial Regimes. *Water International* 44(2): 95–114.

Grönwall, Jenny, and Anna C. Jonsson. 2017. "Regulating Effluents from India's Textile Sector: New Commands and Compliance Monitoring for Zero Liquid Discharge." *Law, Environment and Development Journal* 13(1): 13–31.

Gunderson, L. H., and C. S. Holling. 2002. *Panarchy: Understanding Transformations in Human and Natural Systems*. Washington, DC: Island Press.

Guthman, Julie. 2019. *Wilted Pathogens, Chemicals, and the Fragile Future of the Strawberry Industry*. Oakland: University of California Press.

Halley, D., and C. Shore. 2005. *Corruption: Anthropological Perspectives*. London: Pluto Press.

Haraway, Donna. 1984. "A Cyborg Manifesto Donna Haraway Science, Technology, and Socialist-Feminism in the Late Twentieth Century." In *Simians, Cyborgs and Women: The Reinvention of Nature*, 149–81. New York: Routledge.

Harriss, John. 2007. "Antinomies of Empowerment: Observations on Civil Society, Politics and Urban Governance in India." *Economic and Political Weekly* 42(26): 2716–24.

Hess, David J. 2014. "Smart Meters and Public Acceptance: Comparative Analysis and Governance Implications." *Health, Risk & Society* 16, no. 3: 243–58.

Hess, David J., and Benjamin K. Sovacool. 2020. Sociotechnical Matters: Reviewing and Integrating Science and Technology Studies with Energy Social Science. *Energy Research & Social Science* 65: 101462.

Ho, Selina. 2019. *Thirsty Cities: Social Contracts and Public Goods Provision in China and India*. Cambridge: Cambridge University Press.

Hong, P. Y., N. Al-Jassim, M. I. Ansari, and R. I. Mackie. 2013. "Environmental and Public Health Implications of Water Reuse: Antibiotics, Antibiotic Resistant Bacteria, and Antibiotic Resistance Genes." *Antibiotics* 2: 367–99. doi:10.3390/antibiotics2030367.

Hull, Matthew. 2012. *Government of Paper: The Materiality of Bureaucracy in Urban Pakistan*. Berkeley: University of California Press.

Hurlimann, A., and J. McKay. 2007. "Urban Australians Using Recycled Water for Domestic Non-potable Use: An Evaluation of the Attributes Price, Saltiness, Colour and Odour Using Conjoint Analysis." *Journal of Environmental Management* 83(1): 93–104.

Jamwal, Priyanka, Bejoy K. Thomas, Sharachchandra Lele, and Veena Srinivasan. 2014. "Addressing Water Stress through Wastewater Reuse: Complexities and Challenges in Bangalore, India." Proceedings of the Resilient Cities 2014 Congress. Session G3: Urbanizing Watersheds: A Basin-Level Approach to Water Stress in Developing Cities. https://core.ac.uk/download/pdf/1438 46046.pdf.

Jariwala, C. M. 2000. "The Directions of Environmental Justice: An Overview." In *Fifty Years of the Supreme Court of India*, S. K. Verma and K. Kusum, eds., 469–70. Delhi: Oxford University Press.

Jitendra. 2015. "Will Access to Toilets Guarantee Women's Security in Rural India?" *Down to Earth*, July 4. https://www.downtoearth.org.in/news/will -access-to-toilets-guarantee-womens-security-in-rural-india-44667.

Joshi, S., and G. Shambaugh. 2018. "Oversized Solutions to Big Problems: The Political Economy of Partnerships and Environmental Cleanup in India." *Environment and Development* 28: 3–18.

Kapur, Jagga. 1998. "Preface." In: *Supreme Court on Public Interest Litigation X-XV*, J. Kapur, ed. New Delhi: LIPS Publications.

Karpouzoglou, Timothy, and Anna Zimmer. 2016. Ways of Knowing the Wastewaterscape: Urban Political Ecology and the Politics of Wastewater in Delhi, India. *Habitat International* 30: 1–11.

Kelkar, U. G., and M. N. Thippeswamy. 2012. "Experience of Bangalore in Reuse—Lessons Learned [PowerPoint presentation]." Wastewater Recycle and Reuse: The Need of the Hour Workshop. April 18, 2012. New Delhi: Ministry of Urban Development, Government of India.

Kelkar, U. G., M. Wable, and A. Shukla. 2012. "Valley Integrated Water Resource Management: the Bangalore Experience of Indirect Potable Reuse India-Bangalore." Accessed December 12, 2013. http://www.reclaimedwater.net /data/files/228.pdf.

Kemerink-Seyoum, J. S., T. Chitata, C. Domínguez Guzmán, L. M. Novoa-Sanchez, and M. Z. Zwarteveen. 2019. "Attention to Sociotechnical Tinkering with Irrigation Infrastructure as a Way to Rethink Water Governance." *Water* 11: 1670.

Kochhar R. 2020. "The Virus in the Rivers: Histories and Antibiotic Afterlives of the Bacteriophage at the Sangam in Allahabad." *Notes and Records: Royal Soc.* 74(4): 625–51.

Krishna Chaitanya, S. V., and S. V. Krishna. 2017. "What Drought? Check Out IIT Madras, It's an Oasis." *New Indian Express*, May 11, 2017.

Accessed July 14, 2018. http://www.newindianexpress.com/cities/chennai /2017/may/11/what-drought-check-out-iit-madras-its-an-oasis-1603523--1 .html.

Krishnamurthy, Rohini. 2019. How a Village School in Karnataka Is Tackling Its Wastewater. *Connect*, March 27. https://connect.iisc.ac.in/2019/03/how-a -village-school-in-karnataka-is-tackling-its-wastewater/.

Kumar, Anuj. 2016. National Green Tribunal: A New Mandate towards Protection of Environment. *Legal Desire*. http://www.legaldesire.com/national -green-tribunal-a-new-mandate-towards-protection-of-environment/.

Kundu, Debolina. 2011. "Elite Capture in Participatory Urban Governance." *Economic and Political Weekly* 46(10): 23–25.

Kuttuva, P., S. Lele, and G. V. Mendez. 2018. "Decentralized Wastewater Systems in Bengaluru, India: Success or Failure?" *Water Economics and Policy* 4(2): 1650043.

Lamba, Manisha, and Shaikh Ziauddin Ahammad. 2017. Sewage Treatment Effluents in Delhi: A Key Contributor of B-Lactam Resistant Bacteria and Genes to the Environment. *Chemosphere* 188: 249–56.

Larkin, B. 2013. "The Poetics and Politics of Infrastructure." *Annu. Rev. Anthropol.* 42: 327–43.

Lefebvre, Olivier. 2018. "Beyond NEWater: An Insight into Singapore's Water Reuse Prospects." *Current Opinion in Environmental Science & Health* 2: 26–31.

Lemanski, C., and Tawa Lama-Rewal, S. 2013. "The 'Missing Middle' Class and Urban Governance in Delhi's Unauthorised Colonies." *Transactions of the Institute of British Geographers* 38: 91–105.

Lewis, Michael. 2013. "National Parks, Tiger Reserves and Biosphere Reserves in Independent India." In *Civilizing Nature: National Parks in Global Historical Perspective*, Bernhard Gissibl, Sabine Höhler, and Patrick Kupper, eds., 224–39. New York: Berghahn Books.

Lienhoop Nele, Emad K. Al-Karablieh, Amer Z. Salman, and Jaime A. Cardona. 2014. "Environmental Cost-Benefit Analysis of Decentralised Wastewater Treatment and Re-use: A Case Study of Rural Jordan." *Water Policy* 16: 323–39.

Liu, L., E. Lopez, L. Dueñas-Osorio, L. Stadler, Y. Xie, P. J. J. Alvarez, and Q. Li. 2020. "The Importance of System Configuration for Distributed Direct Potable Water Reuse." *Nature Sustainability* 3: 548–55.

Lloyd Owen, David. 2016. "Public-Private Partnerships in the Water Reuse Sector: A Global Assessment." *International Journal of Water Resources Development* 32: 1–10. 10.1080/07900627.2015.1137211.

Ludden, D. 1996. *Contesting the Nation: Religion, Community, and the Politics of Democracy in India*. Philadelphia: University of Pennsylvania Press.

Mallory, Adrian, Daniel Akrofi, Jenica Dizon, Sourav Mohanty, Alison Parker, Dolores Rey Vicario, Sharada Prasad, Indunee Welivita, Tim Brewer, Sneha

Mekala, Dilshaad Bundhoo, Kenny Lynch, Prajna Mishra, Simon Willcock, and Paul Hutchings. 2020. "Evaluating the Circular Economy for Sanitation: Findings from a Multi-case Approach." *Science of the Total Environment* 744: 140871.

Mathur, N. 2015. *Paper Tiger: Law, Bureaucracy and the Developmental State in Himalayan India.* Delhi: Cambridge University Press.

Maurya, Nutan, Karthick Radhakrishnan, K. Alley, S. Das, and J. Barr. 2017. "A Review Report of the Decentralized Wastewater Treatment System (DEWATS) of Kachhpura Agra." Unpublished report.

McDonald, David A. 2018. "Remunicipalization: The Future of Water Services?" *Geoforum* 91, 47–56.

McGranahan, G. 2015. "Realizing the Right to Sanitation in Deprived Urban Communities: Meeting the Challenges of Collective Action, Coproduction, Affordability, and Housing Tenure." *World Development*, 68, 242–53.

Molinos-Senante, M., F. Hernández-Sancho, and R. Sala-Garrido. 2011. "Cost-Benefit Analysis of Water-Reuse Projects for Environmental Purposes: A Case Study for Spanish Wastewater Treatment Plants." *Journal of Environmental Management* 92(12): 3091–97.

Morales, Margaret C. 2016. My Pipes Say I Am Powerful: Belonging and Class as Constructed through Our Sewers. *WIREs Water* 3: 63–73. doi: 10.1002 /wat2.1108.

Muir, S., and A. Gupta. 2018. "Rethinking the Anthropology of Corruption. An Introduction to Supplement 18." *Current Anthropology* 59, Supplement 18: S4–S15.

Murray-Rust, Dave, Katerina Gorkovenko, Dan Burnett, and Daniel Richards. 2019. "Entangled Ethnography: Towards a Collective Future Understanding." Proceedings of Halfway to the Future Symposium, November 19–20, Nottingham, UK. https://doi.org/10.1145/3363384.336340.

Murty, M. N., and S. Kumar. 2011. Water Pollution in India: An Economic Appraisal. In *India Infrastructure Report*, 285–98. Oxford: Oxford University Press.

O'Donnell, Erin L. 2018. "At the Intersection of the Sacred and the Legal: Rights for Nature in Uttarakhand, India." *Journal of Environmental Law* 30:135–44.

O'Donnell, Erin L., and J. Talbot-Jones. 2018. Creating Legal Rights for Rivers: Lessons from Australia, New Zealand, and India. *Ecology and Society* 23: 7.

Olson, Mancur. 1965. *Logic of Collective Action: Public Goods and the Theory of Groups.* Cambridge, MA: Harvard University Press.

Ostrom, E. 1996. "Crossing the Great Divide: Coproduction, Synergy, and Development." *World Development* 24, 1073–87.

Pandey, Kundan. 2015. "AAP Government Announces Free Water, Cheap Electricity for Delhi Residents." *Down to Earth*, February 25, 2015.

Pasmore, W., C. Francis, J. Haldeman, A. Shani. 1982. "Sociotechnical Systems: A North American Reflection on Empirical Studies of the Seventies." *Human Relations* 35(12): 1179–204. https://doi.org/10.1177/001872678203501207.

Pasmore, William, Stu Winby, Susan Albers Mohrman, and Rick Vanasse. 2019. "Reflections: Sociotechnical Systems Design and Organization Change." *Journal of Change Management* 19(2): 67–85, doi: 10.1080/14697017.2018.1553761.

Peccia, Jordan, Alessandro Zulli, Doug E. Brackney, Nathan D. Grubaugh, Edward H. Kaplan, Arnau Casanovas-Massana, Albert I. Ko, Amyn A. Malik, Dennis Wang, Mike Wang, Daniel M. Weinberger, and Saad B. Omer. 2020. "SARS-CoV-2 RNA Concentrations in Primary Municipal Sewage Sludge as a Leading Indicator of COVID-19 Outbreak Dynamics." *Nature Biotechnology* 38: 1164–67.

Pierce, Gregory, and Silvia R. Gonzalez. 2017. "Public Drinking Water System Coverage and Its Discontents: The Prevalence and Severity of Water Access Problems in California's Mobile Home Parks." *Environmental Justice* 10(5).

Radcliffe, John C., and Declan Page. 2020. "Water Reuse and Recycling in Australia: History, Current Situation and Future Perspectives." *Water Cycle* 1: 19–40.

Randeria, Shalini. 2003a. "Between Cunning States and Unaccountable International Institutions: Social Movements and Rights of Local Communities to Common Property Resources." WZB Discussion Paper SP IV.

———. 2003b. "Glocalization of Law: Environmental Justice, World Bank, NGOs and the Cunning State in India." *Current Sociology* 51(3/4): 305–28.

Rao, Ursula. 2018. Biometric Bodies, Or How to Make Electronic Fingerprinting Work in India. *Body and Society* 24(3): 68–94.

Rath, Manas, Tatjana Schellenberg, Pallavi Rajan, and Geeta Singha. 2020. "Decentralized Wastewater and Fecal Sludge Management: Case Studies from India." *ADBI Development Case Study No. 2020-4* (September). Asian Development Bank Institute.

Ravishankar, Chaya, Sunil Nautiyal, and Manasi Seshaiah. 2018. "Social Acceptance for Reclaimed Water Use: A Case Study in Bengaluru." *Recycling* 3: 4. doi:10.3390/recycling3010004.

Razzaque, Jona. 2004. *Public Interest Environmental Litigation in India, Pakistan, and Bangladesh*, Eric W. Orts and Kurt Deketelaere eds., 21–22. The Hague: Kluwer Law International.

Reno, Joshua. 2014. "Toward a New Theory of Waste: From 'Matter Out of Place' to Signs of Life." *Theory, Culture & Society* 31(6): 3–27.

Resource Recovery and Reuse (RRR) Project. 2012. Bangalore: Institutional Analysis of RRR Business Models.

Rohilla, Suresh Kumar, and Deblina Dwivedi. 2013. *Re-Invent, Recycle and Reuse: Toolkit on Decentralized Wastewater Management*. Delhi: Center for Science and Environment.

Roller, Z., and D. Mayorga. 2017. "An Equitable Water Future: A National Briefing Paper." U.S. Water Alliance, July. https://urbanwaterslearningnetwork.org/wp-content/uploads/2017/07/WaterAlliance_WaterEquity2017.pdf.

Roomratanapun, W. 2001. "Introducing Centralised Wastewater Treatment in Bangkok: A Study of Factors Determining Its Acceptability." *Habitat International* 25(3): 359–71.

Rose, Nikolas. 2007. *The Politics of Life Itself: Biomedicine, Power, and Subjectivity in the Twenty-First Century.* Princeton, NJ: Princeton University Press.

Rosencranz, Armin, Shyam Divan, and Martha L. Noble. 1992. *Environmental Law and Policy in India: Cases, Materials and Statutes.* Bombay: Tripathi Pvt Ltd.

Roy, Ananya. 2005. "Urban Informality: Toward an Epistemology of Planning." *Journal of the American Planning Association* 71(2): 147–58.

———. 2009. "Why India Cannot Plan Its Cities: Informality, Insurgence and the Idiom of Urbanization." *Planning Theory* 8(1): 76–87.

———. 2012. "Urban Informality: The Production of Space and Practice of Planning." In *The Oxford Handbook of Urban Planning*, R. Weber and R. Crane, eds., 691–705. Oxford: Oxford University Press.

Roy, A., and A. Ong, eds. 2012. *Worlding Cities: Asian Experiments and the Art of Being Global.* Malden, MA: Wiley-Blackwell.

Ruiz-Quintanilla, S. Antonio, J. Bunge, A. Freeman-Gallant, and E. Cohen-Rosenthal. 1996. "Employee Participation in Pollution Reduction: A Sociotechnical Perspective." *Bus. Strat. Env.* 5: 137–44.

Sadoway, D., and S. Shekhar. 2014. "(Re)Prioritizing Citizens in Smart Cities Governance: Examples of Smart Citizenship from Urban India." *Journal of Community Informatics* 10(3).

Sagar. 2017. "Down the Drain: How the Swachh Bharat Mission Is Heading for Failure." *The Caravan*, April 30. https://caravanmagazine.in/reportage/swachh-bharat-mission-heading-failure.

Sanghi, Rashmi, ed. 2014. *Our National River Ganga: Lifeline of Millions.* Cham, Switzerland: Springer.

Schindler, Seth, and Federico Demaria. 2016. "'Garbage Is Gold': Waste-Based Commodity Frontiers, Modes of Valorization and Ecological Distribution Conflicts." *Capitalism Nature Socialism* 31(4): 52–59.

Scruggs, C. E., and C. M. Heyne. 2021. Extending Traditional Water Supplies in Inland Communities with Nontraditional Solutions to Water Scarcity. *Wiley Interdisciplinary Reviews: Water*, e1543.

Scruggs, Caroline E., Desmond F. Lawler, George Tchobanoglous, Bruce M. Thomson, Megan R. Schwarzman, Kerry J. Howe, and Andrew J. Schuler. 2020. "Potable Water Reuse in Small Inland Communities: Oasis or Mirage?" *American Water Works Association* 112(4): 10–17.

Sengupta, Sushmita. 2018. "At Least 200 Cities Are Fast Running Out of Water." *Down to Earth*, March 31. http://www.downtoearth.org.in/news/bengaluru

-beijing-mexico-city-and-istanbul-are-some-of-the-cities-that-are-headed
-towards-day-zero-59984#.WrRr_-fLJAE.facebook.

Seth, Bharat Lal. 2012. "Treated Sewage for Gurgaon Projects." *Down to Earth*, August 24. https://www.downtoearth.org.in/news/treated-sewage-for -gurgaon-projects--38955.

Sharma, A. 2018. New Brooms and Old: Sweeping Up Corruption in India, One Law at a Time. *Current Anthropology* 59 (suppl. 18): S72–S82.

Sharma, R. K, M. Yadav, and R. Gupta. 2016. "Water Quality and Sustainability in India: Challenges and Opportunities." In *Chemistry and Water*, S. Ahuja, ed., 183–205. Amsterdam: Elsevier.

Shrotria, Sudha. 2015. "Environmental Justice: Is the National Green Tribunal of India Effective?" *Environmental Law Review*, September 1.

Singh, Bhasha. 2014. *Unseen: The Truth about India's Manual Scavengers*. Delhi: Penguin.

Singh, Gurdip. 1995. *Environmental Law: International and National Perspectives*. New Delhi: Lawman.

Srivastava, Vinay K. 2021. A Grand Privatization Plan to get India's Economy Going. *Mint 11*, March. https://www.livemint.com/opinion/online-views/a -grand-privatization-plan-to-get-india-s-economy-going-11615444707857.html.

Starkl, M., N. Brunner, A. Werner, M. Feil, and H. Kasan. 2018. "Addressing Sustainability of Sanitation Systems: Can It Be Standardized?" *South Africa International Journal of Standardization Research* 16(1).

Starkl, Marcus, Norbert Brunner, and Thor-Axel Stenström. 2017. "Why Do Water and Sanitation Systems for the Poor Still Fail? Policy Analysis in Economically Advanced Developing Countries." *Environmental Science and Technology* 47: 6102–10.

Starkl, Markus, Josephine Anthony, Enrique Aymerich, Norbert Brunner, Caroline Chubilleau, Sukanya Das, Makarand M. Ghangrekar, Absar Ahmad Kazmi, Ligy Philip, and Anju Singh. 2018. "Interpreting Best Available Technologies More Flexibly: A Policy Perspective for Municipal Wastewater Management in India and Other Developing Countries." *Environmental Impact Assessment Review* 71: 132–41.

Strauss, Sarah, Stephanie Rupp, and Thomas Love, eds. 2013. *Cultures of Energy: Power, Practices, Technologies*. New York: Routledge

Suneethi, S., G. Keerthiga, R. Soundhar, M. Kanmani, T. Boobalan, D. Krithika, and L. Philip. 2015. "Qualitative Evaluation of Small-Scale Municipal Wastewater Treatment Plants (WWTPs) in South India." *Water Practice and Technology* 10(4): 711–19.

Sutherland, Catherine, Eva Reynaert, Sifiso Dhlamini, Fanelesibonge Magwaza, Juri Lienert, Michel E. Riechmann, Sibongile Buthelezi, Duduzile Khumalo, Eberhard Morgenroth, Kai M. Udert, and Rebecca C. Sindall. 2021. "Socio-technical Analysis of a Sanitation Innovation in a Peri-urban Household in Durban, South Africa." *Science of the Total Environment* 755, pt. 2: 143284.

Tare, V., and G. Roy. 2015. "The Ganga: A Trickle of Hope." In *Living Rivers, Dying Rivers: A Quest through India*, R. Iyer, ed. New Delhi, India: Oxford University Press.

Truelove, Y. (2011). "(Re-) Conceptualizing Water Inequality in Delhi, India through a Feminist Political Ecology Framework." *Geoforum* 42(2): 143–52.

Urade, Vishal, and Prashant Gondane. 2021. "Effects of the Implementation of Grey Water Reuse Systems on Construction Cost and Project Schedule." *VIVA-Tech International Journal for Research and Innovation* 1(4).

Van der Veer, P. 1994. *Religious Nationalism: Hindus and Muslims in India.* Berkeley: University of California Press.

Von Schnitzler, Antina. 2008. "Citizenship Prepaid: Water, Calculability, and Techno-Politics in South Africa." *Journal of South African Studies* 34(4): 899–917.

Walsh, Casey. 2018. *Virtuous Waters: Mineral Springs, Bathing and Infrastructure in Mexico.* Oakland: University of California Press.

Wanderer, Emily. 2020. *The Life of a Pest: An Ethnography of Biological Invasion in Mexico.* Oakland: University of California Press.

Water Environment Federation, Water Science and Engineering Center. 2019. "Water Science and Engineering Center." Distributed Systems Overview. https://www.wef.org/globalassets/assets-wef/3---resources/topics/a-n/distributed-systems/technical-resources/wsec-2019-fs-012-wef_wrf_distributed_sytems_overview.pdf.

Wells, Joshua. 2014. "Keep Calm and Remain Human: How We Have Always Been Cyborgs and Theories on the Technological Present of Anthropology." *Reviews in Anthropology* 43: 5–34.

Workman, Cassandra L., Maryann R Cairns, Francis L de los Reyes III, and Matthew E Verbyla. 2021. "Global Water, Sanitation, and Hygiene Approaches: Anthropological Contributions and Future Directions for Engineering." *Environmental Engineering Science* 38(5: 402–17.

World Bank. 2017. "Sanitation and Water for All: How Can the Financing Gap Be Filled?" Working paper. March. https://openknowledge.worldbank.org/handle/10986/26458.

Wright, C. Y., L. Godfrey, and G. Armiento. 2019. "Circular Economy and Environmental Health in Low- and Middle-Income Countries." *Global Health* 15: 65.

Yesaya, Mabvuto, and Elizabeth Tilley. 2021. "Sludge Bomb: The Impending Sludge Emptying and Treatment Crisis in Blantyre, Malawi." *Journal of Environmental Management* 277: 111474.

Zhu, Zhongfan, and Jie Dou. 2018. "Current Status of Reclaimed Water in China: An Overview." *Journal of Water Reuse and Desalination* 8(3): 293–307.

Index

Founded in 1893,
UNIVERSITY OF CALIFORNIA PRESS
publishes bold, progressive books and journals
on topics in the arts, humanities, social sciences,
and natural sciences—with a focus on social
justice issues—that inspire thought and action
among readers worldwide.

The UC PRESS FOUNDATION
raises funds to uphold the press's vital role
as an independent, nonprofit publisher, and
receives philanthropic support from a wide
range of individuals and institutions—and from
committed readers like you. To learn more, visit
ucpress.edu/supportus.

www.ingramcontent.com/pod-product-compliance
Lightning Source LLC
Chambersburg PA
CBHW030818270326
41928CB00007B/796